UNDERSTANDING THE LANGUAGE OF MATHEMATICS

A New Common-Sense Method for Learning and
Teaching Mathematics, which Enhances and Liberates
the Brain's Ability to Scientifically Think and Reason

ALEXANDER FIRESTONE

Paperback: 978-1-963883-17-6
eBook: 978-1-963883-18-3
Library of Congress Control Number: 2024904849

Ordering Information:

Prime Seven Media
518 Landmann St.
Tomah City, WI 54660

Printed in the United States of America

Table of Contents

Book Summary

The book is designed like a discourse between a more knowledgeable experienced teacher and the student who wants to improve their confidence and problem-solving ability by learning how to use Mathematics and Mathematical thinking. The book uses English and translates accurately into Mathematics. It introduces the use of Intellectual and Material tools, ideas which are supported by the Theories of Vygotsky an internationally recognized and respected Educational Researcher and psychologist.

The book explains Mathematics, in an easy to read and understanding manner. It offers an easy-to-follow pathway for understanding how to problem-solve and use mathematical thinking, using the language of Mathematics. It's an unusual book as it engages the student who is mathematically challenged, as well as the student who feels very confident in, Mathematics.

It stresses, that students have difficulty because Mathematical ideas are taught in one's native language (in this book English), instead of the Mathematics Language which is a Problem-solving, Logical, Scientific Language that has a completely different language structure than any other language. To translate English into Mathematics you must be bilingual, conversant in both languages. It would be identical to you trying to have a serious conversation with a native French speaker who knows no English, when you don't know any French.

This book is a must read for all Primary School teachers and any other teacher who wants to understand the language of mathematics and wants to learn how to more efficiently be able to problem-solve. This also applies to all parents, students and adults, who would like to be more confident, and capable using Mathematics and Mathematical thinking.

Acknowledgement

This book is dedicated to my mother, Golda Firestone, who worked tirelessly to ensure I received a good education; to the numerous students who encouraged me to write this book and the thousands who had faith in me as a teacher. Also, to my children who persevered and listened to the many hours of my explanations of why things worked the way they did, particularly to Laura, and James, who helped me to more easily explain my ideas and theories, and who spent many hours challenging them; to the many friends and people who persevered all my explanations, thoughts, and theories and to my grandson Justin who was always available to help with the many IT problems I encountered. Lastly to Gloria who volunteered to proofread my writing not realizing how many long hours this would involve.

CHAPTER 1

Mathematics, a Problem-Solving language

A. Introduction to the Language of Mathematics.

Mathematics is unique. It's a Problem-Solving, Logical, Scientific Language and a universal thinking tool. Our brain naturally engages in logical thinking and problem-solving, it's Mathematically thinking all the time, even while you're asleep.

Mathematics is a subject that is in harmony, alignment and complimentary with our thinking processes. However, the irony of the situation, is that Mathematics is considered difficult to understand, appreciate, master and enjoy by many adults and students.

The brain's main thinking function is to make sense of things, so it can use this knowledge and understanding. If a new idea or thought is presented to the brain in a simple, organized, logical, manner, the brain would have no difficulty learning, and assimilating the new idea.

Mathematics is the only language in the world that is taught with little Comprehension and Understanding. Without Comprehension and understanding, a language becomes meaningless sounds and squiggles on paper. This is the main reason that students find Mathematics difficult to: comprehend, understand, learn, and use.

Throughout history, humanity has generated innovative ideas to tackle difficult challenges, resolve conflicts, and craft essential items using only two types of tools. They are either Material, or Intellectual.

1. **Material Tools:** These are tangible objects or procedures using materials like wood, metal, plastic, stone, etc. Examples include hammers, saws, welders, computers, and automobiles. The complete list would be almost endless. While very powerful, they are not always easy to transport, and often require Mental Tools (design drawings), for construction or effective use.

2. **Mental Tools:** These tools influence our thinking, thoughts, ideas, feelings, and behaviour. They include: language, mathematics, symbols, writing, drawings, signs, music, art, dance, body language, sounds, lights, and design drawings. Again the list is endless, but unlike material tools, mental tools are: always available, can be stored in our brains, on paper, or electronic devices, and ready for use throughout our lives.

Mathematics is a language which uses both mental and material tools. For example: Thinking and analysis can be supported by material tools such as pens, pencils, whiteboards, calculators, computers, measuring instruments, microscopes, chemistry, the list continues to increase with new inventions and ideas. The thinking process is enhanced with Concepts and Rules and a language structure where:

- One symbol only has one meaning.
- One word has only one meaning.

- One sentence has only one meaning.
- A paragraph can have two or more meanings, but always the same two or more.

When you're using or learning Mathematics you're working with two languages, your native language and Mathematics. The languages are entirely different, so you'll have to be bilingual, but from the start you've no understanding of the Mathematics language, yet you need to know and understand it to efficiently communicate in it. Imagine the situation of a French person who doesn't know English trying to speak to an English person who doesn't know French. Right from the start of learning Mathematics you're in a similar situation. I can't emphasis enough the need for Mathematics lessons to include learning and understanding the Language to enable students to be successful Mathematics learners.

Understanding Mathematics, like any language, starts with learning how to use it. Comprehension and Understanding are crucial, as without them, what we learn would appear as a collection of meaningless symbols and sounds. Unfortunately, Mathematics is taught with little comprehension, and understanding, and without emphasizing the power that Concepts, Rules, Reflection, and Visualization are, to our thinking and reasoning. This results in many students expressing:

- "I don't understand it."
- "It doesn't make any sense."
- "I can't see any value to it."
- "I'll never use most of this."
- "It's too difficult and complicated to learn."

When students express to others these feelings, they are reinforced by parents and others. I had the same feelings when I studied Mathematics. This then creates the feeling that they are not smart enough to learn Mathematics. This is a tragedy, as it is not lack of intellect that is causing the learning problem, but lack of understanding which is due to the way Mathematics is being taught.

It doesn't make any difference how smart you are, if the subject is taught with little understanding and comprehension, you'll have great difficulty being proficient in the subject. A major problem is that Mathematics is taught using the student's native language which is not logical and problem-solving, and used without the great care needed to translate into Mathematics, where in Mathematics each word and sentence has a specific meaning. This can cause great difficulties and apprehension.

For example: When translating using English: a symbol can have many meanings; a word can have eight or more; a sentence of five words, each word averaged five different meanings can create a sentence with over 3,000 possible translations. Fortunately, the brain can sort very quickly. It would be looking to make sense of the sentence, using common and familiar word meanings that you use first. But the common meanings of the words for the teacher could be different than that of the individual student, and that applies between individual students. In addition, the teacher may not be proficient in the language of Mathematics, especially when it comes to Concepts and Rules, so the risk of student misunderstanding is significant.

Mathematics stands apart as a language, uniquely able to communicate ideas, thoughts, and feelings:

- Clearly
- Correctly
- and Concisely (The three C's)

B. Using Mathematics Symbols

In Mathematics symbols and words have precise meanings:

- One symbol, word, or sentence has only one discrete meaning.
- These meanings are universal, making Mathematics the International Language.
- A paragraph may have multiple meanings, but always the same ones.
- For instructions or definitions, it uses: Concepts which always tells you what to do and Rules which always tells you how to do it, and a Why which causes you to Reflect, and Visualize, on the correctness of Why you're doing it.

There are only two types of symbols. Either:

- Concrete or.
- Abstract.

A Concrete Symbol either:

- Looks like ☐ (any four-sided shape with four 90 degree angles).
- Sounds like. Warning bell or siren etc.
- Feels like. Smooth, rough, bumpy.
- Smells like. Rose or skunk.
- Tastes like. Sweet or sour.

What you're referring to.

Algebraic variables and Algebraic symbols

- These are symbols that you decide to use to represent a word, idea, or mental process. It could be anything.

For example, if you were keeping track of the number of apples being sold to customers, to decrease the amount of information required to remember and work with each time, you would use the symbol A for apple. The symbol by itself is unclear. To eliminate any confusion the symbol would be defined by writing "Let A = apple". Now, until you decide to re-define it, anyone who sees the symbol A, would know we are working with apples. A lot of different words begin with A; apricots, avocados, artichokes, etc. By defining the symbol, you are following the principle "One symbol only one meaning".

- The logical procedure is to use the first letter of the word you'd like to represent. When using abstract symbols, they must always be defined to eliminate confusion. They only have the defined meaning during the time you are using that definition. If you later want to use A for apricot, you must re-define the symbol. Because the meaning of the symbol could change, the Abstract Symbol is called an "algebraic variable". The word variable signals the meaning might be different in a different situation.

Examples of algebraic variables:

- 5A = five apples only when A is defined as A = apple.

- 5A = five apricots only when A is defined as A = apricot

- 5A = five areas only when A is defined as A = area.

- This idea of defining symbols is a pattern which you can use for any situation, for example grocery list: f = fish, m = meat, M = milk etc,

Examples of algebraic symbols

Some symbols have a universal use, for example t = time and T = Temperature. V = volume A = area v = velocity a = acceleration F = force m = mass. When you are working with problems that involve these symbols it is not necessary to define them, they are universally used for communication. Numerals are algebraic symbols that universally only have one meaning. The number "3" is a measure of quantity, three things for example. Alphabet letters are variables in languages. Their pronunciation depends on other letters in the word. In Mathematics there are no exceptions when a symbol is defined it retains that meaning until the definition is changed.

Numerals are algebraic symbols they universally only have one meaning, for example "3" tells you the quantity that you are communicating. If we Let f = fish then 5f means five fish. In English the symbol × means multiply. In mathematics the symbol × means "groups of". Thus 5×3 means five groups of 3; 3+3+3+3+3. The value of 5 × 3 would be communicated as: 5×3=15

> Because their meaning changes according to the situation their meaning must be clearly defined so that they are completely known and understood. The word variable signals to the problem-solver that the symbol's meaning is constant only in the present situation. In a later situation it could have a completely different meaning.

Mathematics is a Problem-solving Logical, Scientific Language.

In Mathematics the three main Mental Tools are: What, How, and Why.

- The Concept - Always tells you What to do. It is the shortest string of information which conveys understanding.

- The Rule - Always tells you How to do it. It is the shortest string of information which conveys understanding.
- The Why - Reminds you to Reflect and Visualize; Why are you using this Concept and Rule, have you achieved what you expected to achieve, it's keeping you on track.
- The Concept and the Rule are the shortest strings of information which carry the understanding, and their understanding is transferable to many different situations.
- The Why looks at the long-term memory and subconscious-memory, looking for patterns of previous like-minded problem-solving situations, positive and negative.
- The Why is a conscious and subconscious habit being formed in problem-solving centred around: Why are you Thinking, Feeling, and behaving this way.

The brain thinks logically, so when problem-solving you must communicate with it logically. When you are solving logical problems there are three words that must come to mind before you even start. The brain needs to know What you're to do, How you're going to do it, and Why you're doing it?

If you are going to work efficiently in any situation, but you are not told specifically What to do; How to do it; and Why you're doing it; how can you possibly think you can efficiently do the task?

Mathematics should be student's favourite subject because Mathematical thinking is helping the brain think better by assisting its ability to think logically. You are supplying the brain with the mental tools, problem-solving experience, and developing a habit that will only get better with time for solving future problem-solving situations.

As a teacher, I have worked with many thousands of students over the last fifty years. I have been told many times to go easy with this class, as they are mostly low-level in ability. I have never found these low-level students. But in almost all my classes I have found students who are severely lacking in the understanding needed for working with basic mathematical principles and procedures, as well as, having little confidence in their ability to think mathematically.

With the methods in this book students will gain the understanding and confidence to become confident efficient problem-solvers. They will also soon see that a mistake is not a failure, but an opportunity to grow intellectually. The mistake has shown them what skills they were deficient in and how to correct the deficiency. **And most important:** It is not how much effort you put into solving a problem, but effort in the correct direction. The more effort you put into problem-solving, where you continually challenge yourself, the more you will grow as a confident, capable problem-solver and thinker.

In this book you will be learning and understanding how to:

- Solve a given problem using Concepts and Rules.
- Gain experience in problem- solving.
- Put ideas together in an organized, logical, scientific, manner.
- Choose the appropriate mental tools, Concepts, and Rules.
- Generalize what you have learned to different problems.
- Build your confidence, capability, and understanding for solving everyday problem-solving challenges.
- Gain a better understanding of why Mathematical Thinking is empowering, important, and essential.

CHAPTER 2

Improving Comprehension and Understanding in Mathematics

Everything in this chapter is a prerequisite in order to completely explain how we use Mathematical instructions in the next chapter "What, How, and Why Thinking".

In Mathematics a single symbol could represent a word, a sentence, a whole paragraph or a procedure. This ability allows you to Chunk information. It facilitates the storing of the symbol's meaning in either the short term, long term, or subconscious memory. It also enables simplicity and conciseness in problem-solving. To make it easier to remember the symbol, the usual practice is to use the first letter, either upper or lower case, of the word or sentence it's representing. For clarity the meaning of the symbol must be defined, before it is used.

In Mathematics the equals ("=") symbol is a Mental Tool. It's a Concept, and an Abstract symbol. It has only One meaning "the left side has the identical value or meaning as the right side or vice versa". In the examples of symbol use below. A = Area means "A" is identical in value to the area on the right.

The English language meanings of "=" are multiple; equals, the same, equivalent, identical etc. A one-dollar coin vending machine defines the cost of what it sells as a one-dollar coin. It has defined $1 as a single $1 coin. You could say two fifty cent pieces, five twenty cent pieces, or ten, ten cent pieces have the identical value of one dollar. But, because they are not matching the vending machine's tariff definition, they are not identical, equal, or the same.

Examples of algebraic symbols which are frequently used in Mathematics are given below. Symbols are very useful, convenient, and easy to use.

Let A = area. Let v = *velocity.* Let V = volume. Let B = size of the triangle's angle. Let b = length of the triangle's opposite side. Let D = diameter. Let d = distance, t = time, T= temperature. The significance of algebra and algebraic symbols as the foundation of Mathematics will be thoroughly explored and clarified as we get closer to solving problems with Equations.

When we try to explain the meaning of very explicit algebraic symbols or statements using English, it's very easy to cause confusion for students. The Algebraic symbol has a very specific meaning, but the English words used to explain the meaning, can individually have eight or more meanings. So, the sentence being used can end up being very misleading, in fact sometimes quite contradictory. It's important that the mathematics meaning is the one followed when working in Mathematics as the Mathematics meaning is explicit.

A command or instruction in Mathematics contains two parts:

- A Concept, which always tells you WHAT TO DO.
- A Rule which always tells you HOW TO DO IT.

- And during problem-solving, Reflection and Visualization on the reasons for and against a particular action. This means it also contains a WHY ARE YOU DOING IT.

The Reflective part involves contemplating:

- Why the chosen Concept and Rule are being employed.
- What outcomes are anticipated.

Visualization entails revisiting past problem-solving experiences stored in the long term or subconscious memory, and looking for similar patterns which can be remodelled to suit the present problem. It also involves asking:

- How the previous experience compares with the present situation.
- What successfully solved similar problems can be recalled.
- Which Mental Tools aided in the prior comparable situation.
- Whether any past experiences can be modified to fit the current situation.

These structural differences distinguish Mathematics from the English language, necessitating a UNIQUE approach when translating from English into Mathematics. This is not the case when translating from Mathematics to English, quite the contrary. This is why you must sometimes do the problem first, to understand the English translation. Traditionally word problems in Mathematics are considered difficult. You'll come to realize as you work through this book, IT'S NOT THE MATHEMATICS THAT'S DIFFICULT, but understanding the English translation of, or into, Mathematics.

The skills you're learning in Mathematics, such as problem-solving, comprehension, technique, and understanding, can be applied to any logical problem. Solving math problems is just a small part of the whole subject. What's crucial is that from solving mathematics problems, where everything is well structured and logical, you're gaining the ability to generalize problem-solving experiences, which are applicable for solving logical problems for out-side the classroom. These are compelling reasons for everyone to be competent in mathematical thinking. You've already learned that when there is no space between symbols it means the symbols are multiplied together. For example, if A = apples then:

> 6A means six apples. This is more convenient for communication than
>
> 6 × A or, A+A+A+A+A+A for six apples.

The procedures mentioned above are easy to learn. By solving problems using the methods taught in this book you will be gaining a better understanding of Mathematics and problem-solving. These techniques will become second nature in your everyday problem-solving experiences.

To increase the versatility of symbols we use subscripts and superscripts.

1. **Subscripts:** Frequently when we are working with symbols, we need to communicate more information about what the main symbol stands for. We do this with a subscript. The subscript is a smaller font symbol which is written in front of the symbol, at the bottom, with no space between it and the main symbol. This small font symbol gives you more information.

    Example 1: In a shop we are selling either apples or bananas. They differ in colour. Either red, yellow, or green. It's necessary to keep track of what we have sold. In this situation a subscript will help.

Procedure:

- The NUMBER of items in the quantity is written first,
- Second with no space between them, the SYMBOL which represents what we are selling,
- third with no space between, and at the bottom, the SUBSCRIPT, which is in a smaller font (size).

Given: We have sold 24 red apples, 15 green apples, and 35 yellow apples.

WE MUST FIRST DEFINE THE SYMBOLS BEING USED:

Let A = apple, R = red, Y = Yellow, and G = green.

We would mathematically write this as: $24A_R + 15A_G + 35A_Y$

We must also be able to communicate what we've sold. Remember the brain works logically, it's always trying to make sense of what you have written. Would you please translate what the algebraic expression "$24A_R + 15A_G + 35A_Y$ communicates. Say it softly to yourself." Please say it again.

Did you read the above expression as 24 apples that are red, 15 apples that are green and 35 apples that are yellow? If you did you were following what you've always been told since Primary. "You always read or write from left to right".

If you started with 24, paused for an instant and then quickly said Red apples plus 15 Green apples, plus 35 yellow apples, you have demonstrated you are a quick learner and a good logical thinker. You went against everything you've been told and read it logically instead. In Mathematics language we translate, and communicate, logically. That means clearly, correctly, and concisely. In Mathematics you can read from left to right or right to left when you are doing multiplication or addition. When it's subtraction it is always left to right. For division you must follow the correct convention which is indicated in the problem. You'll learn more about division during the Chapters on Division and Fractions. In the Chapter on Fractions you'll see the power of reading the idea of a sentence Right to Left, and Left to Right when you study the Three One Rules. They're probably among the most powerful Mental Tools in Mathematics

Example 2: Tomorrow in the shop you're also selling flowers, some of them are purple, others white. You need three new symbols to add to the previous list. F = flowers, W = white, and P = purple. Below is what you sold today:

$64A_R + 32A_G + 17A_Y + 30F_Y + 26F_P + 24F_R + 15F_W$

Did you read this using your mathematical thinking as: Sixty-four red apples, Thirty-two green apples, Seventeen yellow apples, Thirty yellow flowers, Twenty-six purple flowers, Twenty-four red flowers, and Fifteen white flowers? If you did you are already learning ideas that will help you outside the classroom.

With more practice you will automatically read left to right then right to left, or whichever way makes the reading clearer.

Frequently you only need to communicate a list of algebraic variables. Below are two examples of how you would do this.

Given:

$64A_R + 32A_G + 14A_Y + 7F_P$ you would say 64 A sub R plus 32 A sub G plus 14 A sub Y plus 7 F sub P

This would allow the receiver of what you've just communicated to accurately duplicate the algebraic expression for his own use.

Given an expression you are not familiar with, and not knowing the meanings of the symbols, you would still be able to communicate it.

For example. Given: you would say

$12x_2 + 7x_3 + 5y_2 + 9y_3$ you would say

12x sub 2 plus 7 x sub 3 + 5y sub 2 + 9y sub 3

2. **Superscripts**: To explain superscripts, we first must define multiplication which is a mathematics instruction.

Multiplication means multiple addition. The grouping symbol "×" is used for this.

For example, if we want to multiply 6 by 5 we would write this instruction as 5 × 6 = 5 groups of 6 = 6 + 6 + 6 + 6 + 6 (REMEMBER "=" means identical in value or meaning)

- 5 × 6 = 6 + 6 + 6 +6 + 6 which means five groups of six.
- 14 × 6 = 6 + 6 + 6 + 6 + 6 + 6 + 6 + 6 +6 + 6 + 6 + 6 + 6 + 6
- 6 ×17 = 17 + 17 + 17 + 17 + 17 + 17

The multiplication symbol means "multiple addition". "Groups of" is a clearer definition so "groups of" is the mathematical meaning that we'll use. Frequently it is necessary to multiply a number by itself several times. The superscript (Index) symbol is used for this instruction. A small font symbol, an Index, is written in front of the symbol, without space and at the top. For example:

Five multiplied by itself three times would be written as:

5^3 which $= 5 \times 5 \times 5$. We say "five to the index three". Seven to the index four means $7^4 = 7 \times 7 \times 7 \times 7$

$8^1 = 8$ Eight to the Index 1

The index (superscript) and subscript are not always numbers, they could also be algebraic variables. For example: Given $2^A = 16$ determine the value of A

$2^A = 16 = 4 \times 4 = 2 \times 2 \times 2 \times 2 = 2^4$ We discover, by definition of the "=" Concept, that,

$2^A = 2^4$ The left side of the = is identical in value on the right side.

This can only be true if $A = 4$. This is logical thinking and the skill which you will be using throughout the book.

Chapter 1. and 2. Have given you enough Mathematics Comprehension and Understanding to enable you to start putting ideas together in an organized, logical, scientific way. Chapter 3 will also help you consolidate what you have already learned. A problem-solving book is planned with tailored practice, support, examples, and challenging problems to augment this first book. In the meantime any school, or library has all sorts of Mathematics books which give you problems you can practice on with the answers in the book. Please remember the answer book should only be used to check your understanding.

Careless mistakes should always be found when you check your solution procedure. In your everyday activities there is neither an answer book, nor a solution manual to do every problem. When you have determined the correct solution, you must act on it. If you weren't proficient and careful in your thinking the consequence of a mistake could be significant.

You are practicing and practicing, building your confidence, and competence, so that you always know What you're doing, How you're going to do it, and Why you have used the Concepts and Rules you have. **Remember you Have the ability, it's been the lack of understanding which has cause you difficulty. Finishing this book will prove that to you.** Completing this book is only the start of your journey. It's up to you, whether you want to gain the skills you need, to reach what you're aiming for, in life. You may not be aiming to major in Mathematics. You are working to become a confident, capable, problem-solver, who understands the language of Mathematics. You are developing the skills needed to be successful in any field that requires Thinking and Analytical skills. You would of course be capable of becoming an excellent Mathematics teacher.

CHAPTER 3

What, How, and Why Thinking. Understanding The Number System

Problem-solving in Mathematics, or working with logical problems, means working with intellectual tools, which are facts. Your brain is excellent for problem-solving, thinking logically and scientifically. But, there must be a good reason for your thinking, feeling and behaviour. **If it's not a good reason; Why not? What is your reason?** Your thinking and reasoning **must be morally driven**. That's the type of thinking needed for creating a happy, well-functioning relationship, family, community or society. In this book you're gaining more practice, working as partners and friend with your brain, you're learning to work together as one. You are not just memorizing a procedure, which doesn't make any sense to the brain, only enabling you, to gain a good result, on a school Assessment. You are supplying your brain with the mental tools, and knowledge it needs to make better decisions.

The main intellectual tools used in Mathematics are Concepts, Rules, and Why, these are facts. Some of the Mathematical meanings for Concepts, Rules, and Why, were given in Chapter 1 and 2. It's important now, to get practice in using them during problem-solving. This is the only way you can feel confident and competent while using the Model in Fig 1 Below.

What, How, and Why, Model

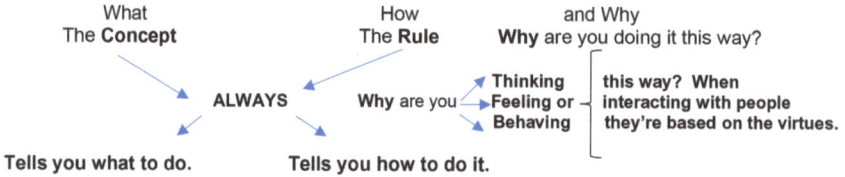

What	How	and Why
The **Concept**	The **Rule**	**Why** are you doing it this way?

ALWAYS   Why are you → Thinking / Feeling or / Behaving → this way? When interacting with people they're based on the virtues.

Tells you what to do.   Tells you how to do it.

Fig. 1

A. The Number System. Working with Concept, Rule, and Why.

All of the Commands and instructions in Mathematics, as well as in logical problems follow the above Model in Fig. 1. When you are thinking Logically and scientifically, one idea naturally follows from the previous idea. You are using facts to analyse a situation and arriving at an answer which makes sense and is reliable.

Chapter 1 and 2 provides the knowledge and understanding for us to start using the "What, How, and Why" model of thinking to develop the Concept and Rule for counting, and the creation of the Number System.

If you have difficulty following this learning approach. No problem, logical thinking is difficult at first. You're developing a new habit of knowing the reason why you're doing what you're doing. This new habit is important for you. You might need to return, or jump to ideas and understanding from previous chapters, or those following Chapter 3. When you do this, don't bog yourself down, TAKE ONLY WHAT YOU NEED to understand this Chapter, then return. Rome wasn't built in a day. Go slowly, you will go faster with time as you gain confidence and familiarity with the ideas. It's not ability that you lack, only previous understanding and knowledge.

For complete understanding of any topic, you need ideas, knowledge, and understanding, from many places, and directions. You'll gain more information, experience, and understanding, from weaving together the new knowledge and understanding with the old knowledge and understanding that you already have.

Our Number System started with counting and keeping track of what we had. We did this using a one-to-one relationship with our fingers. If we had three goats and had to keep track of them as they grazed in the fields, we hold up three fingers, and in a one-to-one relationship we matched a finger to a goat. This system worked well but was deficient, as the number of goats you had to look after changed you had to remember how many fingers you used. The problem was solved by using pebbles, kept in a pouch, instead of remembering the number of fingers. Of course, other methods were used, slashes on a stick, knots in a string etc. When you started out each morning, the number of goats you started with was recorded by Pebbles in a bowl. They could easily be added or taken out as the number of things changed.

Three goats ➡ three fingers ➡ three pebbles   ➡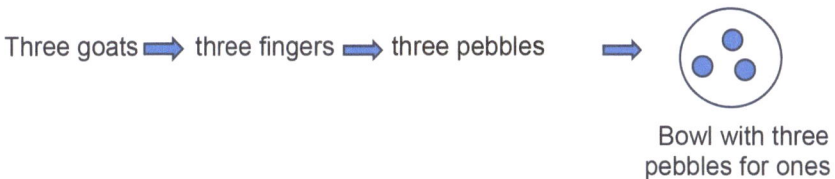

Bowl with three
pebbles for ones

There was still a problem. How many pebbles were in the bowl had to be physically shown. How could you tell someone how many pebbles were in the bowl?

Creating a common language was more difficult, but eventually names were created and accepted for the numbers: one one, two ones, three ones, four ones, five ones, six ones, seven ones, eight ones, and nine ones, and their accompanying symbols: 1, 2, 3, 4, 5, 6, 7, 8, and 9. This was a big improvement, but meant there was more to remember than just counting pebbles by looking at them. If the number of pebbles was more than 9 you needed a bigger bowl and it was more difficult to count them.

Eureka! A new idea. Everyone has ten fingers. They could use that number as a package size - a package of ten. We could use bigger pebbles for packages of ten and smaller pebbles for the ones. What we needed now was a symbol for ten. Another new idea!

There was no reason to use only one bowl. We could use one bowl for the packages of Ten the other for the Ones. This was a good idea. **Eureka!** We could always put the bowls in the same order. The first bowl on the right for the "ones" and the second bowl on the left for the "tens". We still have a problem. What symbol can we use to indicate the bowl is empty? What does the bowl look like when its empty? A circle. This makes sense. The circle indicates the bowl is empty meaning there are no pebbles in the bowl. **Eureka!** If we had ten pebbles we would represent the ten as 10. One in the ten bowl and zero in the ones bowl.

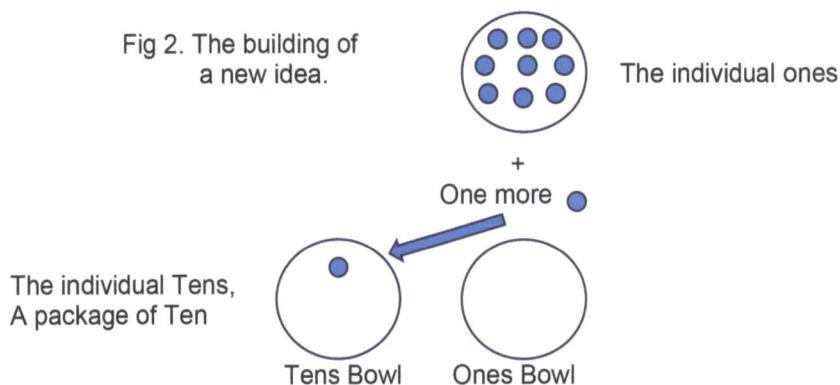

Fig 2. The building of
a new idea.

The individual ones

+

One more

The individual Tens,
A package of Ten

Tens Bowl   Ones Bowl

Eureka! Keep the bowls in the same order. The Ones on the right, the smaller size. The Tens on the left the larger size. We have already determined that the symbol for ten will be 10. Therefore 10 indicates that there is one package of ten and no ones .

For reinforcement if we had one pebble in the tens bowl and none in the ones bowl, it made sense to write 10. If we had three tens and four ones, we could write 34. Everything started to make sense, and the new system was easy to learn. This convinced every one of the importance of learning the new symbols. It was easier to count pebbles in the bowl, but not good enough when the number came larger. The newer language was creating a common language which was easy to learn and understand. In this new system the first symbol was always size One and the number indicated how many Ones. The second number indicated size Ten and how many Tens. This pattern was consistent, robust, logical and easy to learn. It's a Rule. This meant that there now was a system for accurately communicating larger quantities.

In this new system the first symbol was always size One and the number indicated how many Ones. The second number indicated size Ten and how many Tens. This pattern was consistent, robust, logical and easy to learn. It's a Rule. This meant that there now was a system for accurately communicating larger quantities.

In this new system the first symbol was always size One and the number indicated how many Ones. The second number indicated size Ten and how many Tens. This pattern was consistent, robust, logical and easy to learn. It's a Rule. This meant that there now was a system for accurately communicating larger quantities.

For example: 25 represents two Tens and five Ones.
                 74 represents seven Tens and four Ones.
                 80 represents eight Tens and no Ones.
                 14 represents one Ten and four Ones

When the quantity was larger than 90, a third size was created and called the Hundreds. We now have:

369 represent three Hundred, six tens and nine Ones.
207 represent two Hundred, no tens and seven Ones.

Later when quantities became larger new names were agreed upon for the New larger sizes, but the same Number pattern was followed.

Ten Hundred became a Thousand. Ten Thousand became size Ten Thousand.

Ten Ten thousand became a Hundred Thousand, and ten Hundred Thousand became size Million.

A separate change was made when the sizes reached the Thousands, the names of the Ones when being used to count the number of Tens changed to make it easier to communicate. The new names are as follows:

11 One ten and One became eleven
12 One ten and two became twelve

A new pattern was used for the rest of the Tens and Ones. The ones were modified so they ended with teen:

13 became thirteen
14 became fourteen
15 became fifteen
16 became sixteen
17 became seventeen
18 became eighteen
19 became nineteen

The Tens became: twenty, thirty, forty, fifty, sixty, seventy, eighty, ninety.

The pattern for the numbers became:

> 42 became forty-two
> 75 became seventy-five and so on.
> 367 became three hundred and sixty-seven
> 894 became eight hundred and ninety-four
> 714 became seven hundred and fourteen
> 3, 526 became three thousand, five hundred and twenty-six
> 7,611 became seven thousand six hundred and eleven
> And so on. Only the comma was used to separate the number into groups of three digits to make it easier to recognize the larger sizes.
> 13,425,117 became thirteen million, four hundred and twenty five thousand, one hundred and seventeen.

One unfortunate consequence of changing the names of the Ones and Tens is that the Number System was straying away from the logical reasoning that developed the new Number System. When children are first learning the number system, they are learning how easy it is to learn, when things make sense. With only needing to know the nine single digits, the empty symbol, packages of ten and the three names for the sizes, they could easily count to 999. However, students must continue with the logic of the original number system and say, '9 hundreds, 9 tens, and 9 ones.

This author believes that the beginning learning system should be taught until students become conversant with the logic behind the Number system. At the age of two and a half children can learn how to count up to 9 Hundred, 9 Tens, and 9 Ones and enjoy the beauty of putting ideas together in a organized, logical, scientific manner. Once they have reached this stage of seeing the beauty of thinking logically, there is no harm to them seeing, that for convenience, we'll change how we communicate using the new naming system.

When sizes less than one occurred, another change was necessary. The same Number Pattern was kept. Each size was still a factor of ten larger than the previous size. All that was needed was a full stop "." to separate the whole sizes from the new fractional sizes. The first size less than one, the Ones were divided into ten equal parts and each part was called the Tenths. For the next smaller size, the Tenths were divided into ten equal smaller parts. This next smaller size was called the hundredths. For the third the Hundredths were divided up by ten to make the thousandths, and so on.

The pattern for the numbers became:

0.1.1   represents no Ones and one Tenth

0.8   represents no Ones and eight Tenths

0.46   represents no Ones and forty-six Hundredths

0.367   represents no Ones, three Hundreds and sixty-seven Thousandths

23.62   represents twenty-three and sixty-two Hundredths

6.234   represents six and two hundred and thirty-four Thousandths

17.456 represents seventeen and four hundred and fifty-six Thousandths

This is how humanity progresses, sharing ideas through language and putting them together in an organized, logical, scientific way, by using Mental Tools.

Initial Concept: We use our fingers to count things

Initial Rule: We use a one-to-one relationship with our fingers and what we are counting.

B. From empty bowl to the Base 10 Number System.

What are we doing with a quantity so that we can communicate the quantity? Arrange the quantity into different sizes groups.

Number Concept – Arrange the quantity into simple sizes.
Number Rules –

1. The first whole number size is called the Ones.
2. Each next larger size increases by a factor equal to the Base Number.
3. The maximum unit in any size is equal to one less than the Base Number.
4. The sizes increase going from left to right.
5. The symbol 0 indicates the size is empty.
6. For Base sizes other than 10, a small subscript is used for indicating the Base Size (32_4 indicates a Base four number)

For ease of communication, we can use our index knowledge. Each size increases by a factor of 10. $10^0 = 1$; $10^1 = 10$; $10^2 = 100$; $10^3 = 1,000$ and so on.
For Example: A Base 10 Number
Each size increases by a factor of 10 (The Base Number) going from right to left.

Given 3,964.735. It is composed of the following sizes

Thousands	Hundreds	Tens	Ones	Tenths	Hundredths	Thousandths
$\times 10^3$	$\times 10^2$	$\times 10^1$	$\times 10^0$	$\times 10^{-1}$	$\times 10^{-2}$	$\times 10^{-3}$
3,000	**900**	**60**	**4**	**0.7**	**0.03**	**0.005**
3×10^3	9×10^2	6×10^1	4×10^0	7×10^{-1}	3×10^{-2}	5×10^{-3}

It's important that you become familiar with and understand all of the above ways of describing a quantity. They allow you to easily work with very large or small numbers. In our scientific world you will come across all of the above.

It is now possible to formulate a simple Concept and Rule system for Number. Each size in any Number System is based on a package size equal to the Base of the Number System

The Universal Number System

Number Concept – Arrange the quantity into groups based on the Base Size.

Number Rules:

1. The first whole number size is the ONES.
2. Each following larger size is increased by a factor, equal to the Base of Number. Size 10 = Ten × Size ONES'. Size One = equal to Ten × Size Tenths etc.
3. A decimal point separates the "ONES" from the fractional sizes.
4. The sizes increase going from right to left.
5. The maximum number in anyone size is ALWAYS one less than the Base size.
6. The symbol "zero" signifies the size is empty.
7. If the Base is not 10 a small subscript is used to identify the different Base.

This Same Pattern with modification; can be used in any recurring pattern.

For example: In a factory, a product can be sold as a single small bottle, a small box of four bottles, a package of four small boxes, a carton of six packages, a pallet of 24 cartons, or a Truck load of 18 pallets. Knowing this Number system, you could easily determine how many bottles, small boxes, or cartons etc. are required to fill large or small orders. You receive an order for a full truck load. You must supply the Preparation dept. the correct number of bottles, small boxes, packages, cartons, and pallets for them to complete the order. How many bottles would you need for 7 packages.

A problem using this information will be demonstrated on pg.8, Chapter 8 Conversions.

Given the number 33022_4 What does this mean? Answer it's a Base 4. Number, so starting with the 2 on the Right:

33022_4 represents $2 \times 4^0 + 2 \times 4^1 + 0 \times 4^2 + 3 \times 4^3 + 3 \times 4^4$

Converting each size to a Base 4 number value then a Base 10 Number value you'd have:

$2 \times 4^0 = 2 \times 1 = 2$	2
$2 \times 4^1 = 2 \times 4 = 8$	8
$0 \times 4^2 = 0 \times 4 \times 4 = 0$	0
$3 \times 4^3 = 3 \times 4 \times 4 \times 4 = 3 \times 64 = 192$	1 9 2
$3 \times 4^4 = 3 \times 4 \times 4 \times 4 \times 4 = 3 \times 256 = 768$	7 6 8
	<u>1 2</u>
Value as a Base 10 Number	9 7 0

33022_4 = Base 10 = $9 \times 10^2 + 7 \times 10^1 + 0 \times 10^0 = 970$

You are now starting to understand the Base 10 Number System. You realize that you are working with factors of 10:

Base 10
- $1 = 1 \times 10^0 = 1$
- $10 = 1 \times 10^1 = 10$
- $100 = 10 \times 10 = 1 \times 10^2 = 100$
- $1,000 = 10 \times 10 \times 10 = 1 \times 10^3 = 1,000$
- $10,000 = 10 \times 10 \times 10 \times 10 = 10^4 = 10,000$

- $6 = 6 \times 10^0 = 6$
- $60 = 6 \times 10^1 = 60$
- $600 = 6 \times 10^2 = 6 \times 10 \times 10 = 600$
- $6,000 = 6 \times 10^3 = 6 \times 10 \times 10 \times 10 = 6,000$
- $60,000 = 6 \times 10^4 = 6 \times 10 \times 10 \times 10 \times 10 = 60,000$

Base 4
- $1 = 4^0$.
- $4 = 4^1$
- $16 = 4 \times 4 = 4^2$
- $64 = 4 \times 4 \times 4 = 4^3$
- $256 = 4 \times 4 \times 4 \times 4 = 4^4$

Index of Ten Rule: × 10^x

We can see from the Base 10 Number System each time you multiply by 10 the size becomes an index of I greater. This is a definite pattern and is reproducible. We can take advantage of this pattern, using the Index 10 Notation:

Multiplying by 10 results in moving the decimal point one size to the left. The value of the Index tells you the number of places to move, right for a positive index, left for a negative index.

> 1000 can be represented as 1 move the decimal 3 places to the right

> .0001 can be represented as 1 move the decimal 3 places to the right

We can represent this idea with a new Mental Tool. See Fig. 3 below.

Signs for direction to move are: $^+$ to right — negative to left   Number of places to move

$$\times 10^{-4}$$

Groups of 10 Move the decimal

Fig. 3

For examples:

(a) $500 \times 10^{-6} = 0.0005$
(b) $5 \times 10^{-3} = 0.005$.
(c) $0.0005 \times 10^{-6} = 0.0000000005$
(d) $42 \times 10^{-6} \times 2 \times 10^{-3} \ 3 \times 10^{10} = 252 \times 10 = 2520$

Remember the Concepts and Rules are facts and ALWAYS tell you What to do and How to do it. They are reliable mental tools. The tools you need in problem-solving. We'll wait until further in the book to determine mathematically the value of 10^0 . We don't have enough experience and mental tools yet to completely prove the Index of Ten system. We are assuming the value to the Index Zero of the numbers we have used so far is always 1. We'll have to wait until pages 158-159, Chapter 8. New RULE to prove $10^0 = 1$, and when N represents any number $N^0 = 1$. In the meantime, we use the present Rule as though it's a fact and always true.

We'll start a new investigation Counting with pebbles. Let's have another think about it

Let's use our grouping symbol and make a table:

We should be able to see there are dependable. re-occurring Rules here:

1 × 1 = ●   "One multiplied by any value becomes that value"
2 × 1 = ● + ●
3 × 1 = ● + ● + ●
4 × 1 = ● + ● + ● + ●
5 × 1 = ● + ● + ● + ● + ●

        "Any value multiplied by One remains the same in value"

1 × 2 = ● + ●
1 × 3 = ● + ● + ●
1 × 4 = ● + ● + ● + ●
1 × 5 = ● + ● + ● + ● + ●

We can definitely see a pattern here:

If you have one group of 1 pebble you end up with 1 pebble.
If you have one group of 2 pebbles, you end up with 2 pebbles.
If you have one group of 3 pebbles, you end up with 3 pebbles.
If you have one group of 4 pebbles, you end up with 4 pebbles, and so on.
We can also see:

If you have one group of 1 pebble you end up with 1 pebble.
If you have two groups of pebbles, you end up with 2 pebbles.
If you have one group of 3 pebbles, you end up with 3 pebbles.
If you have one group of 4 pebbles, you end up with 4 pebbles, and so on

If you had 3 groups of one pebble, you end up with three pebbles.
If you had 4 groups of one pebble, you end up with four pebbles, and so on.

What we have done is started an intellectual journey by exploring a simple idea.

$$1 \times 1 = 1 \quad \text{and} \quad 1 \times 1 = 1$$
$$2 \times 1 = 2 \quad \text{and} \quad 1 \times 2 = 2$$
$$3 \times 1 = 3 \quad \text{and} \quad 1 \times 3 = 3$$
$$4 \times 1 = 4 \quad \text{and} \quad 1 \times 4 = 4$$

From this observation and resultant pattern we can Formalize a Rule.

Any Number multiplied by 1 stays the same in value and
1 multiplied by any Number stays the same in value

The idea above is a very simple observation, we can also in the situation above: when multiplying by 1 you can read from left to right or right to left. Does this work for any number situation? We'll reinvestigate this question in Chapter 6 as well as continuing with our investigation of the simple idea. It will lead you to the formation of **The Three One Rules. They are among the most powerful problem-solving tools** for helping to discover new ideas, solving Equations and Fraction problems. These Rules will help you understand the above section as well as give you a glimpse of the Rules power, and importance in problem-solving. I would recommend you jumping ahead to Chapter 6. Read only as much as you understand, then return to where you are now. **You'll see that even a simple idea can hold the gemstone of a very powerful mental tool,** for helping to discover new ideas, solving Equations and Fraction problems. Chapter 6 will help increase understanding in some of the other sections in Chapter 3 as well as sections in other Chapters.

Let's get some more experience in Number Systems. What is the meaning of 7654? It's the value of a quantity expressed in the Base 10 Number System. Given that $10^0 = 1$ (We will derive the value of any number to the index 0 in Chapter 6) change 7654 to Index Form.

$$4 \times 10^0 = 4 \times 1 \qquad\qquad = \qquad 4$$
$$5 \times 10 = 5 \times 10 \qquad\qquad = \qquad 50$$
$$6 \times 10^2 = 6 \times 10 \times 10 \qquad = \qquad 600$$
$$7 \times 10^3 = 7 \times 10 \times 10 \times 10 = \underline{\quad 7000}$$
$$\text{Total} \quad = \quad 7654$$

Given the quantity 21120_3. Translate the value of this Number into a Base 10 Number System.

We notice the subscript of 3 so we know that we are working with a quantity expressed in groups of three. To understand what that quantity is in a Base 10 system we first convert each size quantity into Base 3.

Base 3 – Index Notation Arrange the quantity 21120_3 into groups based on the Base Size (packages of 3) and convert same into Base 10.

We have in Index form;   Base 10 Equivalents

$0 \times 3^0 = 0 \times 1 \qquad\qquad = 0 \times 1 \quad = \qquad 0$

$2 \times 3^1 = 2 \times 3 \qquad\qquad = 2 \times 3 \quad = \qquad 6$

$1 \times 3^2 = 1 \times 3 \times 3 \qquad\quad = 1 \times 9 \quad = \qquad 9$

$1 \times 3^3 = 1 \times 3 \times 3 \times 3 \qquad = 1 \times 27 = \qquad 27$

$2 \times 3^4 = 2 \times 3 \times 3 \times 3 \times 3 = 2 \times 81 = \underline{\quad 162}$

Adding the obtained Base 3 values, $\qquad 12$

we have: $\qquad\qquad 21120_3 \qquad = \underline{204}$ As a Base 10 Number

We will derive the meaning of any number to the index zero when we are working with equations in Chapter 6.

Different Number systems.

In computer systems you can operate in base 2, 4, 8, etc. If you were required to convert the Base 10 number 897 to Base 4. How would you do this?

First set up your Base 4 system as a Table of values of Base 4 with their equivalent Base 10 values. You could use Index notation just like was done with Base 10. When you've done that convert the Base 4 values into Base 10 values. For example:

Base 4 indices   Equivalent Base 10 values

- $4^0 =$ $=$ 1
- $4^1 =$ $=$ 4
- $4^2 = 4 \times 4$ $=$ 1 6
- $4^3 = 4 \times 4 \times 4$ $=$ 6 4
- $4^4 = 4 \times 4 \times 4 \times 4$ $=$ 2 5 6
- $4^5 = 4 \times 4 \times 4 \times 4 \times 4$ $=$ 1 0 2 4

You've got the equivalent Base 10 values, now find the largest size Base 4 size which has a Base 10 value less than 897. From above it can be seen that $4^4 = 256$ is the largest Base 4 size less than 897. W can see that because $4^5 = 1024$ which is larger than 897. We check to see how many groups of 256 the value 897 contains. Using multiple addition, you find :

3×256 is less than 897. For example: $3 \times 256 = 7\ 6\ 8$ This tells us we have 3×4^4 in our Base 4 number. We continue with the rest of the number.

Subtract to see how much of our number is left $3 \times 256 =$
$$\begin{array}{r} 8\ 9\ 7 \\ -\ 7\ 6\ 8 \\ \hline 1\ 2\ 9 \end{array}$$

Repeat the previous procedure. Using your Base 4 Table. What's the largest index 4 size which has a Base 10 value less than 129? Ans. $4^3 = 64$. How many 64 does 129 contain Ans. $2 \times 64 = 128$. We have found that our number contains 2×4^3

Subtracting 128 from 129

we are left with

Any number to index zero is $= 1 = 1 \times 4^0$

$$\begin{array}{r} 1\ 2\ 9 \\ -\ 1\ 2\ 8 \\ \hline 1 \\ -\ 1 \\ \hline 0 \end{array}$$

This demonstrates that the Base 4 number:

$$4^4 \ 4^3 \ 4^2 \ 4^1 \ 4^0$$
$$3 \ 2 \ 0 \ 0 \ 1_4 = \text{Base 10 number 897}$$

Study the above solution procedure until you understand Why you did each step in the solution process. You can set up your own practice problems. You can always check your answer by reversing the procedure. For example, in the above checking your answer procedure is as follows:

$$4^4 \ 4^3 \ 4^2 \ 4^1 \ 4^0$$

Starting with $3 \ 2 \ 0 \ 0 \ 1_4$ translates to. $3 \ 2 \ 0 \ 0 \ 1_4$

$$3 \times 4^4 = 3 \times 256 \ = \ 768$$
$$2 \times 4^3 = 2 \times 64 \ \ = \ 128$$
$$1 \times 4^0 = 1 \times 1 \ \ \ \ = \ \ \ \ \underline{1}$$
$$\text{Total} \ \ \ \ 897$$

Setting up your own problems and then checking your answer actually enables empowerment. By checking your answer you will:

- come up with a different problem-solving procedure which also consists of a number of sophisticated steps
- be getting more practice putting ideas together in a meaningful logical manner
- be very interested in proving that your original answer is correct, which means the work you are doing is meaningful
- believe your first answer was correct, so you came up with ideas more quickly and efficiently
- be solidifying your understanding of the importance of knowing why you are Thinking, Feeling and Behaving this way
- Because the problems are more abstract, and difficult, be building more self-confidence in your problem-solving ability

Let's try one more. Change 342_{30} into a Base 2 Number. All computers are based on a Base 2 system, so, it would be good to compare a larger Base with a smaller one. Like every problem we start with what we know to eventually understand what we didn't know. We are familiar with Base 10 numbers, but we don't know the Base 30 numbers. We need to know the Base 30 values so we can calculate the value of 342_{30}. We can then go from a Base 30 to a Base 10. Then change the obtained Base 10 number and change to a Base 2. We'll first construct a table of the Base 30 numbers with their equivalent Base 10 values.

Base 30 indices.   Equivalent Base10 values

$1 \times 30^0 = 1 \times 1$ $\qquad = \qquad 1$

$1 \times 30^1 = 1 \times 30$ $\qquad = \qquad 30$

$1 \times 30^2 = 1 \times 30 \times 30 \ = \qquad 900$

We now have enough information to construct the Base 30 Number. 342_{30} means

$2 \times 30^0 = 2 \times 1 \qquad = \qquad 2 \qquad\qquad\qquad 2$

$4 \times 30^1 = 4 \times 30 \qquad = \qquad 120 \qquad\qquad 1\ 2\ 0$

$3 \times 30^2 = 3 \times 900 \ = \qquad 2700 \qquad\quad \underline{2\ 7\ 0\ 0}$

$\qquad\qquad\qquad\qquad\qquad\qquad\qquad\qquad\qquad 2\ 8\ 2\ 2$

We now determine the Base 2 Number which is equal in value to the Base 10 Number 2822

To change 2,822 into a Base 2 number we first have to construct the Table of indices for Base 2

$2^0 = 1$ $\qquad\qquad\qquad\qquad 2^6 = 2\times2\times2\times2\times2\times2 = 64$

$2^1 = 2$ $\qquad\qquad\qquad\qquad 2^7 = 2\times2\times2\times2\times2\times2\times2 = 128$

$2^2 = 2\times2 = 4$ $\qquad\qquad 2^8 = 2\times2\times2\times2\times2\times2\times2\times2 = 256$

$2^3 = 2\times2\times2 = 8$ $\qquad 2^9 = 2\times2\times2\times2\times2\times2\times2\times2\times2 = 512$

$2^4 = 2\times2\times2\times2 = 16$ $\qquad 2^{10} = 2\times2\times2\times2\times2\times2\times2\times2\times2\times2 = 1024$

$2^5 = 2\times2\times2\times2\times2 = 32$ $\quad 2^{11} = 2\times2\times2\times2\times2\times2\times2\times2\times2\times2\times2 = 2048$

$2^6 = 2\times2\times2\times2\times2\times2 = 64$

As before we must look for the largest Base 2 value just less than 2,822. From our Base 2 Indices Table we find that 2^{11} which is 2,048 is the largest value less than 2822. We now check to see how many groups of 2,048 does our 2,822 contain. It only contains 1 group. We subtract to see how much of the number we have left.

$$342_{30} = 2,822$$
$$\text{Subtracting} \quad 2^{11} = -2,048$$
$$\overline{774}$$

As before we check to find the largest Base 2 value less than 774. We see that $2^9 = 512$ is the smallest value that 774 contains, and that it only contains1 group of 512. We subtract the 512 to see what we have left of the number.

$$\begin{array}{r} 774 \\ - 512 \\ \hline 262 \end{array}$$

Again we check to find the largest Base 2 value less than the 2 6 2 which is what we have left. We see that is $2^5 = 256$.

We subtract the 256 from what we had left and we have:

$$\begin{array}{r} 262 \\ - 256 \\ \hline 6 \end{array}$$

$2^2 = 4$ is the largest value smaller than 6, and it only contains one group of 2^2. Subtracting we have:

$$\begin{array}{r} 6 \\ - 4 \\ \hline 2 \end{array}$$

We only have a value of 2^1 left.   2

Subtracting $2^1 = 2$ we have $-$ 2

                          0

We subtract and we have nothing left.

We have determined that:

$$2^{11}\ 2^{10}\ 2^9\ 2^8\ 2^7\ 2^6\ 2^5\ 2^4\ 2^3\ 2^2\ 2^1\ 2^0$$
$$342_{30} = 1\ \ 0\ \ 1\ 0\ 0\ 0\ 1\ 0\ 0\ 1\ 1\ 0$$

What have we learned from this? We've discovered a robust pattern which enables us to represent any value number with a Base 2 number. The pattern is very simple 0 or 1. But this pattern could represent several situations. On or Off, positive or negative, light or dark, flowing or not flowing, the valve is open the valve is closed. **Eureka!** When the light is on, electricity is flowing, when the light is off the electricity is off.

This idea was used with the first computers. They had a series of mechanical switches which switched lights on and off. Each light represented a particular index of two. it wasn't long before they remembered an electronic switch, the vacuum diode. It conducted electricity in only one direction. It was a very practical and important switch and used in the first very powerful computer systems. This idea spawned the semiconductor.

The rest is history. Semiconductor diodes could be layered together to make many switches in a very small size. From this simple idea of investigating and refining our understanding of Number systems, using problem-solving, logical thinking and scientific method, using the idea of a malleable pattern super computers were developed. Today they are able to make more than 1,000,000,000,000,000 (a thousand trillion) calculations a second enabling scientists to explore, invent, investigate ideas, and situations that we never knew, existed, and so much yet to be discovered. This is only a glimpse of the power of Mathematical Thinking and what the brain is capable of.

Let's try another Conversion. Change 29,631 to a Base 6 Number. Do it yourself first. Then see how it is done below. We'll follow the same procedure as we need the indices for Base 6 converted to Base 10.

$$
\begin{array}{rcrcr}
1 \times 6^0 & = & 1 & = & 1 \\
1 \times 6^1 & = & 6 & = & 6 \\
1 \times 6^2 & = & 6 \times 6 & = & 3\,6 \\
1 \times 6^3 & = & 6 \times 6 \times 6 & = & 2\,1\,6 \\
1 \times 6^4 & = & 6 \times 6 \times 6 \times 6 & = & 1,2\,9\,6 \\
1 \times 6^5 & = & 6 \times 6 \times 6 \times 6 \times 6 & = & 7,7\,7\,6 \\
1 \times 6^6 & = & 6 \times 6 \times 6 \times 6 \times 6 \times 6 & = & 4\,6,6\,5\,6 \\
\end{array}
$$

First, we need the largest indices of 6 which is smaller than 29,631

29,631 contains groups of 1×6^5. 1×6^5 is equal to 7 7 7 6. 1×6^6 is too large We need to determine how many groups of 1×6^5 there are in 29,631

We find there are three groups $3 \times 6^5 = 3 \times 7\,7\,7\,6 = 23{,}328$.

We have used up 23,328 of our 29,631. We have to subtract to see what we have left.

$$
\begin{array}{r}
2\,9,6\,3\,1 \\
-\ 2\,3,3\,2\,8 \\
\hline
6,3\,0\,3 \\
\end{array}
$$

From our grouping Table we see that 6 3 0 3 contains groups of $1 \times 6^4 = 1296$ We determine how many groups of 1296? A quick use of mental observation sees that it contains 4 groups.

$4 \times 6^4 = 4 \times 1296 = 5184$

We have used up 5184 of our number we must subtract from 6303 to see how much is left.

$$\begin{array}{r} \scriptstyle 9\ 10 \\ \scriptstyle 2\ \cancel{10} \\ 6\,3\,0\,3 \\ -\ 5\,1\,8\,4 \\ \hline 1\,1\,1\,9 \end{array}$$ contains groups of 216 = 1 × 6³

1 1 1 9 contain **5 × 6³** = 5 × 216 = 1080

We must subtract the 1080 from the 1119 to see what is left:

$$\begin{array}{r} 1\,1\,1\,9 \\ -\ 1\,0\,8\,0 \\ \hline 3\,9 \end{array}$$ contains 1 group of 36 = 1 × 6²

39 contains 1 **× 6²** we must subtract from what we have: 39 – 36 = 3

We have **0 × 6¹**
3 contains 3×1
3 = **3 × 6⁰**

We have now determined all of the indices of 6 that 29631 contains and can now display 29631 as a Base 6 Number.

29,631 = 3×6⁵ + 4×6⁴ + 5×6³ + 1×6² + 0×6¹ 3×6⁰
29,631 = 345103₆

We can always check our answer as shown below:

3×6⁰ = 3 × 1	=		3
0×6¹ = 0 × 6	=		0
1×6² = 1 × 3 6	=		3 6
5×6³ = 5 × 2 1 6	=		1, 0 8 0
4×6⁴ = 4 × 1 2 9 6	=		5, 1 8 4
3×6⁵ = 3 × 7 7 7 6	=		2 3, 3 2 8

$$\begin{array}{r} \scriptstyle 2\ 2 \\ \hline 2\,9,\,6\,3\,1 \end{array}$$

You now know that you can exchange one Base number for another when required and can always check your answer. You can set your own Homework problems and do as many as you believe are necessary to gain the skill of working in Mathematics. Solving a few difficult problems always helps you more than doing simple or identical ones. There are many Mathematics Workbooks available in stores to purchase or to borrow in the library. It is important that you remember that understanding will give you the confidence to solve any of your everyday Mathematics problems. The techniques that you will mostly learn in these books are usually only rote learning of procedures that work for the problems presented. If you don't understand what you are doing, you will gain very little understanding and self-confidence from doing more and more similar problems. You will only have been memorizing a procedure and doing the memorized procedure faster.

Below is another practice problem. You should be able to do it by yourself. If not, it's done for you. Make sure that you understand every step along the way.

$3\ 2\ 1\ 0\ 0\ 1_4$ Convert this number to a Base 10

First, we have to organize one-to-one. each Base 4 size to its equivalent value in Base Ten.

$3\ 2\ 1\ 0\ 0\ 1_4$ Remember sizes increase going from right to left and 1 is the smallest whole number.

Our first size $1 \times 4^0 = 1 \times 1 = 1$

$1 \times 4^0 = 1 \times 1$ $= 1$ $=$ 1

$0 \times 4^1 = 0 \times 4$ $= 0$ $=$ 0

$0 \times 4^2 = 0 \times 4 \times 4$ $= 0 \times 16$ $=$ 0

$1 \times 4^3 = 1 \times 4 \times 4 \times 4$ $= 1 \times 64$ $=$ 64

$2 \times 4^4 = 2 \times 4 \times 4 \times 4 \times 4$ $= 2 \times 256$ $=$ 512

$3 \times 4^5 = 3 \times 4 \times 4 \times 4 \times 4 \times 4$ $= 3 \times 1024$ $=$ $\underline{3072}$

                                                          $\overset{1}{}$

Answer $3\overset{}{6}49$

3 6 4 9 is a Base 10 system number, no subscript is necessary.

As you work through the conversion problem, you must verify that the What, How, and Why, thinking, reasoning and procedure, makes sense. This is a habit you're trying to form during problem-solving.

CHAPTER 4

Understanding the Addition Concept and Rule.

We have mentioned that logical problem-solving revolves around three words:

- What, **the Concept, knowing what to do.**
- How, the **Rule, knowing how to do it and**
- **Why,** the **validation of your procedure through reflection and visualization.**

Let's use as an example the simple command "addition". The English translation would be to add, plus, combine, etc. If we are following the mathematics symbol "+" we must use the mathematics meaning. You knew the mathematics meaning when you were about three years old. You learned it from either Mum or Dad. They didn't teach it to you as a Concept and Rule, they didn't know it as such. They only knew how to use it.

For example:

I want you to imagine you are only three years old again playing on the living room floor with your toys. Mum walks in carrying a large open top box and from mum's expression it looks heavy, and you can guess Mum has been shopping. She plonks it down on the edge of the table. You're naturally curious, so what would you do? ....

You'll run over to see what Mum has in the box. You look in the box and see Apples, Oranges and Bananas. Mum says to you, "I'm very tired dear, could you please help me by making a pile of bananas on the corner of the table".

You're eager to help Mum as she wants to have a pile of bananas on the corner of the table. You've only been told "what to do" and "why to do it", does she have to tell you how to do it? Most three-year-old would be able to make a pile of bananas on the corner of the table.

Would you put any apples in the pile? How about putting oranges in the pile? No, Mum said a pile of bananas. At the age of three you already knew that to make a pile of like things you must only put like things together.

This means that you already knew how to mathematically solve an addition problem. Unfortunately, by the time you go to school what you previously learned about addition is now only in your subconscious memory. It's harder to retrieve subconscious information. To retrieve that knowledge, you usually need to be in a situation of stress like the situation you were in when you learned it. Below is the mathematics meaning of the "+" symbol. You follow the Addition Concept and Addition Rule in any mathematics situation where addition is involved:

Addition Concept – Make groups of Like Things.

Addition Rule – Only put Like Things together.

Equals Rule – Equivalents can always be exchanged for equivalents

At school you didn't learn the mathematics meaning of the + sign. You learned to use the English meaning of the "+" symbol. Addition translated to: plus, add, put together, addition etc. In Mathematics a symbol only has one discrete universal meaning, a word only has one meaning, and a sentence only has one meaning. It's confusing for the brain to interpret what it needs to do when it encounters the plus sign. It therefore follows a rote memory procedure which was taught in the classroom. When the brain knows the mathematics Concept, Make groups of Like Things, it knows exactly what to do in any situation and "Starts looking for Like Things".

Let's see if the Concept and Rule are in your subconscious memory. Imagine during the school holidays, you are going for an interview at a small shop, for a part-time job. You know, to get the job, you will have to be able to correctly answer any questions the boss may ask you. You enter the shop. The boss is sitting behind a desk and behind him is a shelf on which there are three transparent bags containing chocolates. The one on the left has two chocolates, the middle one has three chocolates and the one on the right has four chocolates. This is the same as any, 2 + 3 + 4 situation. The boss asks the following questions (you'll find the expected answers you'll be required to give at the end of the problem):

1. "How many bags are on the shelf"? You answer .... He then asks,
2. "How many chocolates are on the shelf"? You answer .... He then asks,
3. "How many bags contain an even number of chocolates"? You answer .... He then asks,
4. "How many bags contain four chocolates"? You answer .... He then asks,
5. "How many bags contain an odd number of chocolates"? You answer .... He then asks,

6. "How many bags contain a prime number of chocolates"? You answer .... He then asks,
7. "How many bags contain a perfect square number of chocolates"? You answer ....

Answers: (1) Three bags are on the shelf.
(2) A total of 9 chocolates are on the shelf.
(3) Two bags contain an even number, one with two chocolates and the other with four chocolates.
(4) One bag contains four chocolates.
(5) One bag contains an odd number of chocolates. The bag with three chocolates.
(6) One bag contains a prime number of chocolates, the three-chocolate bag.
(7) One bag contains a perfect square number of chocolates, the four-chocolate bag.

Notice for the same situation, three bags of chocolates on the shelf, you used your understanding of Concept and Rule, and the Equals Rule to formulate seven different correct answers. We could have given more. For example: How many bags had less than three chocolates, etc. Understanding in mathematics is very important. It will increase your problem-solving confidence, your ability to think and reason, and lead you to new areas of knowledge, understanding, and wisdom.

If you answered all the questions correctly, it was because you were following the Addition Concept and Addition Rule. Also notice all your answers were Clear, Correct, and Concise. There would be no doubt in the boss's mind, that you were a person who knew what to do, how to do it, and could be competent in any addition situation.

Let us now look at an identical type of addition problem given in the normal classroom.

Simplify: 2 + 3 + 4 = ………. I'm guessing your answer was 9

That answer was the English translation of "+" **BUT** the problem is a Mathematics problem.

Mathematically the problem was asking you to **make groups of Like things**. It didn't tell you what the like things you were to make a group of. **Subconsciously** you remembered that 2 stood for two ones, 3 meant three ones, and 4 stood for four ones **so you thought 9 ones** but knew **(rote memory)** the correct classroom answer was just nine. You also knew Subconsciously that 2, 3, and 4 are numbers. Outside the classroom you could have said three numbers, and that would have been correct, but rote memory told you that in the classroom there's only one answer. You've never been given that three numbers was a correct answer. **Surprisingly even at school you don't always communicate the value of a number when communicating.** If the password for your computer was 4321, would you tell someone your password was four thousand three hundred and twenty-one? **I don't think so!** You would have given an easier answer to remember, which had nothing to do with the value of the number, such as 43, 21.

Let's try to reinforce what you have learned so far with another simple problem. You've all seen the ATM machine, in fact many of you have already used one. You know in order to withdraw money, you need a credit/debit card as well as the four digits password to get money out. You know if you don't know these two things the ATM machine won't give you the money. One evening Mum is very tired and not feeling very well. She asks you do a very important errand for her. She asks you, "Do you remember how to use the ATM machine?". You know the ATM is only three blocks away, and you've been with her only a few times but you're very familiar with the procedure. She wants you to withdraw $60 from the machine. There's very little traffic, and there's a bicycle lane to the shop. She wants you to come straight back so that she doesn't have to worry. She gives you the card and, as you watch, writes a password 4 3 5 on a small piece of paper and says: "You can ride your bicycle, your skateboard, or run to the shop, just be very careful. Don't show anyone the password and come straight back". She folds the paper with the password in half, gives it to you and says "Be careful and come right back". What procedure are you going to follow so that you will be back home quickly, safely, and with the $60? Remember, you're now in a problem-solving situation and the task is important. Have you had enough time to think about how you would carry out the task.?

. . . .

The best answer to the above is given right after the **Addition Rule** which is a few lines down the page.

Whenever you see the "+" sign, and you are doing mathematics it's telling you to:

- **Addition Concept - Make groups of Like Things**, and the only way to do this is,

- **Addition Rule - Only put Like Things together.**

Answer to the ATM machine errand. I hope you didn't say I'll take the bicycle or skateboard. You're using What, How, and Why, thinking. Every problem-solving situation starts with knowing What to do, and How to do it.

What are the first things you needed for the task? Do you have the card and the correct password? You've got the card and you've seen the password, but Mum only wrote three numbers. You know it must be four numbers. You should have said to: "Mum, you've only written down three numbers". Mum would have apologized and written down the correct four-digit number 4357. Would she have said the code is four thousand, three hundred, and fifty-seven? I don't think so! We are not always interested in the value of a number. Sometimes we're only interested in the exact number of digits and the order in which they're spoken or written. Do you know the value of the digits that are in your computer password, or your telephone number? I hope the above example will help you remember the importance of knowing the Concept and Rule when problem-solving.

Solving problems with Concepts and Rules is a very powerful, useful, and important skill, that you'll learn by solving your Mathematics problems with Concepts and Rules. Mathematics is working with ideas. Working with numbers is only a very small part of Mathematics.

When you know the Concept and Rule you ALWAYS know exactly what to do there is no guess work and you have full confidence in your answer. Outside the classroom and inside the classroom you should be deciding what Like Things you should put together. This opens the door to all sorts of possibilities. You are not just doing things in a wrote manner but understanding and thinking out the best answer for the situation!

$$\bigcirc + \triangle + \triangle + \triangle + \square + \square + \bigcirc + \square + \bigcirc + \square + \triangle + \bigcirc + \bigcirc + \diagdown\square$$

In the situation above, the brain recognizes the addition instruction. Can you come up with at least eight answers that fit the Concept? At the bottom of the page, I have given you more than 12 answers, (there are still more), that fit the Concept and Rule. Write your answers down on a separate piece of paper. You will see the list of answers which are supplied further down.

You will not come up with a new idea or thought, by just doing things the same old way, or by following a memorized procedure. For a better idea you must think outside the common square. This can only be better done, with understanding and pattern recognition. We will go into much more detail with What, How, and Why later when we are problem-solving, with Equations. First, we must know the fundamentals of addition, subtraction, multiplication and division.

I strongly believe what you are learning at school should be preparing you for your future life adventures in problem-solving. Sadly, in many school situations, much of the learning is only preparing you for the classroom test!

Answers for the shape problem:

How many: shapes (14), different shapes (4), different shapes with corners (4), round shapes (5), square shapes (4), triangular shapes (4), shapes with corners (9), Geometric shapes (14), not geometric(0), not round (9), not triangular(10), not square(9), are closed shapes(14), with one flat side 0, with two flat sides (1), with three flat sides (4), with four flat sides (5)

Typical Addition problems: In each case you are using the Addition Concept and Rule, and the Equals Rule

a)
$$789765$$
$$869878$$
$$+678759$$
$$\overline{2\ 2\ 2\ 2\ 2\ 2}$$
$$2338402$$

b)

week	day	hour	minute	second

$$3wk + 5dy + 7hr + 9min + 29sec$$
$$5wk + 6dy + 9hr + 8min + 27sec$$
$$+\ 6wk + 9dy + 8hr + 7min + 25sec$$

5wk	1 dy		1 min	

$$19wk \quad 0dy \quad 0hr\ 25min \quad 21sec$$

The two problems a) and b) above have been done for you. I have not shown all the steps used to obtain the answers. That's for you to do. You need to remember you are always exchanging equals for equals. When you have more of one item than the size allows you make packages of that smaller item and add the larger packages to the next larger size. For example, in a) 22 ones becomes 2 groups of Ten and 2 groups of One. In b) 81 seconds becomes 1 minute and 21 seconds.

CHAPTER 5

Understanding the Subtraction Concept and Rule.

**The Concept and Rule for Subtraction
and the Why are you using them.**

Warm up ideas. Consider the following two situations:

1. If you owed $223 and had a $55 reduction of debt, How, would you solve this subtraction problem?

2. If you had $223 and you spent $55, how would you solve this subtraction problem?

The topic of Subtraction will be used to introduce the **Concept, Rule, and Why** family in an organized, logical, scientific, way to help develop problem-solving ability.

If you were given the information "−$6" without any further information, it could mean: Either you owe six dollars, or you are taking six dollars away. It's for this reason the Subtraction Concept and Rule states:

The Subtraction Concept - You are taking away from WHAT you OWE or from WHAT you have.

The Subtraction Rule -You are taking Like from Like.

The **Subtraction Concept** symbol "−" means "**what you owe, or what you take away**". The meaning determines how you approach the problem.

The **Concept ALWAYS** tells you **WHAT** to do, the **Rule ALWAYS** tells you **HOW** to do it, the **Why** causes you to Reflect, and Visualize, on why you've chosen the Concepts and Rules you have. They are the basic Mental Tools used in problem-solving.

The specific mental tools required in all Subtraction problems are as follows:

The **Subtraction Concept and Rule, the Why, and Equivalence Rule.**

Equivalence Rule – Equivalents of identical value can be exchanged with each other.

Hardly a day goes by without you using the **Equivalence Rule** in one way or another. First when a young child, when you "Traded" your toy, for someone else's toy. Much later you realized: Two 50 cent pieces for a dollar coin. One gram of water for one cubic millilitre of water. Five one-dollar coins for a $5 note. A Size TEN for ten Size ONES. 60 SECONDS for one MINUTE. 7 days for one WEEK. One day's WORK for $20. 250 ml for one CUP OF LIQUID. The possibilities are numerous.

All of the Subtraction Problems below are solved with four main ideas; What, How, and Why and the Equivalence Rule.

I have given a wide variety of examples and Example solutions following the procedure of What, How, and Why. **When we are exchanging in Mathematics, we are using The Equivalents**

Rule "Equivalent things of identical value can be exchanged for each other".

Back to the warmup: You owe $223 and make a payment of $89. How much do you now owe? In this problem you will be practicing the use of What, How, and Why thinking during the entire problem-solving procedure.

Procedure: Arrange the given information, logically, correctly, clearly, and for ease of solution.

The problem solution begins by knowing What to do. This is a subtraction problem. You'll need the Subtraction Concept and Rule, Why and the Equivalence Rule.

Subtraction Concept – You are taking away from WHAT you OWE or from WHAT you have.

Subtraction Rule – You can only take like from like,

And the **Equivalence Rule** – Equivalents of equal value can be exchange with each other.

In this problem you will be taking away from what you owe. What you owe will have to be on the top and your payment which you will have to take away must be on the bottom.

$$
\begin{array}{r}
- \ \$2\,2\,3 \\
\underline{\$ \ \ 8\,5} \\
\end{array}
$$

You arrange the given information, logically, correctly, and clearly, for ease of use. You make sure that the sizes are aligned so that you will be taking Like from Like. ALWAYS start with the smallest size first, in this case the size Ones. You can see that you can't take 5 ones from 3 ones. You'll have to exchange equivalents for equivalents. You exchange 1 of size Ten for an equivalent number of Ones. **You don't borrow**. Borrowing is not correct and is not logical. There is no confusion to the brain when you follow "Equivalent is exchanged for Equivalent". 1 Ten =10 Ones

$$
\begin{array}{r}
{}^{1\ 10} \\
-\ \$2\,2\!\!\!/\,3 \\
\$\ \ 8\,5 \\
\hline
8
\end{array}
$$

The 2 2 3 represents 2 x 100 + 2 x 10 + 3 x 1. In the problem you are taking 5 Ones from 3 Ones. There aren't enough Ones to subtract 5. We can get more Ones by exchanging 1 of the Tens from the 2 Tens we have. We now have 10 Ones and 3 Ones. We can take the 5 Ones away from the 10 Ones leaving 5 Ones. We now have 5 Ones and 3 Ones giving us 8 Ones. We now recognise a pattern that we can use for the rest of the problem. That is, we are always taking a single digit away from Ten.

By always taking away from Ten we are taking advantage of our knowledge of taking a single digit from Ten and adding the remainder to what we initially had. This is logical and useful knowledge and understanding that we will always use in a Base 10 Number system.

The remainder of 8 Ones must be placed in the Ones position on the final remainder row. You're finished with the Ones, you now work on the Tens.

You are in a similar situation as previously when working with the Ones, except that you now have to take 8 Tens away from the remaining 1 Ten. You'll have to exchange one of the Hundreds for an equivalent number of Tens. 1 Hundred = 10 Tens. This will give you the following on the left.

$$
\begin{array}{r}
{}^{10} \\
{}^{1\ 1\ 10} \\
-\ \$2\!\!\!/\,2\!\!\!/\,3 \\
\$\ \ 8\,5 \\
-\ \$1\,3\,8
\end{array}
$$

As before you subtract the 8 Tens from the exchanged 10 Tens giving you a remainder of 2 Tens. This is added to the 1 Ten that you started with. 2 Tens + 1 Ten = 3 Tens. The 3 Tens is your final remainder for the Tens and it must be placed in the Tens position on the final remainder row at the bottom. There are no Hundreds to take away which leaves you with a final result of – $1 3 8 which you still owe.

+ 1 = 3. This is the remainder you have for the Tens so it goes in the bottom final remainder row.

This is the pattern you follow for all subtraction problems. The pattern can always be modified to fit the particular problem. Let's look at a similar problem.

You have $223 in your account and you make a purchase of $85. After the purchase what will you have left in your account. This time you are taking away from what you have. You arrange the given information, logically, correctly, clearly, and for ease of use. You make sure that the sizes are aligned so that you will be taking Like from Like. ALWAYS start with the smallest size first, in this case the size ONES. This time you are taking away from what you have. You arrange the information as below:

$$\begin{array}{r} \$\ 2\ 2\ 3 \\ -\ \$\ \ \ 8\ 5 \\ \hline \$\ 1\ 3\ 8 \end{array}$$

You follow the same pattern as used in the previous problem. The only change is that you are taking away from what you have, and you end up with a positive answer.

Exercises: The first two problems I have given thousands of times to primary, secondary as well as tertiary students. Have a go at them first without seeing the solution. When they don't follow the Subtraction Pattern that has been demonstrated above, they end up using rote method procedures they learned when they first started subtraction; they end up with the **incorrect answer** and have strong feelings that their answer is correct. For some it is only when they use the calculator, that they realize and believe it's incorrect. This is why there is repetition on some of the ideas I have shared. It's very difficult to break away from a procedure which is simple, only learned in a rote manner, and only works on problems very similar to the ones they have practiced on and solve. Below are 7 problems for practice. Remember you are exchanging equivalents for equivalents, you are not borrowing. **Solutions next page.**

Problem 1. Simplify (Problem solutions are on the next page):

$$
\begin{array}{r}
7\ 6\ 8\ 6\ 7\ 5\ 8\ 0\ 5\ 5 \\
-\ 7\ 8\ 5\ 4\ 3\ 2\ 5\ 1\ 3\ 2 \\
\hline
\end{array}
$$

Problem 2. Simplify:

$$
\begin{array}{rlllll}
& 5\text{wk} & 3\text{dy} & 6\text{hr.} & 7\text{min.} & 8\text{sec} \\
- & 2\text{wk} & 6\text{dy} & 7\text{hr.} & 8\text{min.} & 9\text{sec} \\
\hline
\end{array}
$$

Problem 3. Simplify:

$$
\begin{array}{rl}
& 5.\ 3\ 8\ 7\ 5\text{km} \\
-\ 25.\ & 2\ 0\ 0\ 4\text{m} \\
\hline
& \text{km}
\end{array}
$$

Problem 4. Simplify:

$$
\begin{array}{rlllll}
& 2\text{wk} & 6\text{dy} & 7\text{hr} & 8\text{min} & 9\text{sec} \\
- & 5\text{wk} & 3\text{dy} & 6\text{hr} & 7\text{min} & 8\text{sec} \\
\hline
\end{array}
$$

Problem 5. Simplify:

$$
\begin{array}{r}
6\ 3.\ 1\ 1\ 3\ 2 \\
-\ 6\ 4.\ 0\ 0\ 2\ 1 \\
\hline
\end{array}
$$

Problem 6. Simplify:

Subtract - 5 weeks. 6 days, 4 hours, 6 minutes, 8 seconds, from 7 weeks, 4 days, 3 hours, 2 minutes, 3 seconds. Given 1 wk = 7 dy; 1 dy = 24 hr; 1 hr = 60 min; and 1 min = 60 sec

Problem 7. Simplify:

$$7\ 7\ 9\ 2\ 8\ 6\ 7\ 6\ 9\ 5\ 3\ 7\ 6\ 2\ 8\ 9\ 5\ 4$$
$$-\ 9\ 5\ 2\ 3\ 5\ 2\ 4\ 7\ 9\ 3\ 4\ 2\ 3\ 1\ 7\ 3\ 4\ 3$$

ALWAYS RE-check your work to catch simple mistakes. You are checking your own solution, only to check for understanding. This is not to give busy work, embarrass, or make you feel silly. On the contrary it's demonstrating **how very important**, it is that your Thinking, Feeling, and Behaviour in problem-solving follows known Concepts and Rules and your thinking is dependable.

You should always check your work to make sure your understanding is up to scratch and your answer is dependable. This is when you spot mistakes. When you leave school there's no answer book! You'll have to act on the answers that you have. If they aren't correct the consequence can be horrific.

While solving the problems below, follow the previous Subtraction Pattern.

The pattern demonstrates understanding, and logical reasoning, during each step towards the solution.

Problem 1. Simplify:

$$7\ 6\ 8\ 6\ 7\ 5\ 8\ 0\ 5\ 5$$
$$-\ 7\ 8\ 5\ 4\ 3\ 2\ 5\ 1\ 3\ 2$$

	10	10	10	10			10		
7	4	3	2	1	10	0	2	10	

$$-\ 7\ \cancel{8}\ \cancel{5}\ \cancel{4}\ \cancel{3}\ \cancel{2}\ 5\ \cancel{1}\ \cancel{3}\ 2$$
$$\underline{7\ 6\ 8\ 6\ 7\ 5\ 8\ 0\ 5\ 5}$$
$$-\ 1\ 6\ 7\ 5\ 6\ 7\ 0\ 7\ 7$$

This is a problem where you are taking away from what you owe. The bottom line is larger than the top line. First reverse the two quantities, you are set up to take Like from Like. Follow the previous pattern. You don't have enough Ones to take 5 from 2, so you exchange the next size Ten for ten Ones. This is placed above your existing 2. You are left with only 2 Tens which is not enough to take 5 away from. You exchange 1×100 for 10×10. The 10×10 are placed above the 2 in the size Ten column. This procedure continues as you continue the subtraction process. Your final result is: − 1 6 7 5 6 7 0 7 7

Your final result is what is left over after you have subtracted.

You can always check your answer by adding your result to what you took away. You will then get back to the number you started with. For example:

$$
\begin{array}{r}
7\ 6\ 8\ 6\ 7\ 5\ 8\ 0\ 5\ 5 \\
+\quad 1\ 6\ 7\ 5\ 6\ 7\ 0\ 7\ 7 \\
\hline
\scriptstyle 1\ 1\ 1\ 1\ 1\quad\ 1\ 1 \\
7\ 8\ 5\ 4\ 3\ 2\ 5\ 1\ 3\ 2
\end{array}
$$

Which is what you started with

Problem 2. Simplify:

	7dy	24hr	60min	
	2dy	5hr	6min	
4wk				60sec
5wk	3dy	6hr	7min	8sec
− 2wk	6dy	7hr	8min	9sec
2wk	3dy	22hr	58min	59sec

The problem is asking you to take away from what you have. It's set up to take Like from Like. Everything is straight forward but remember that you're exchanging equivalents for equivalents You would take 9sec from 60sec. The 51sec remainder would be added to the 8sec that you started with, this gives you 59sec remainder. You would repeat this process for the other sizes.

Problem 3. Simplify:

$$
\begin{array}{r}
5.\ 3\ 8\ 7\ 5\text{km} \\
-\ 25.\ 2\ 0\ 0\ 4\text{m} \\
\hline
\text{km.}
\end{array}
$$

You are not set up for taking Like from Like. The Remainder is to be in km. first change m into km. into km. m = .001km. therefore 2 5. 2 0 0 4m = 0. 0 2 5 2 0 0 4km

$$
\begin{array}{r}
9\ \ 9 \\
4\ \ \cancel{10}\ \cancel{10}\ 10 \\
5.\ 3\ 8\ 7\ \cancel{5}\ 0\ 0\ 0\text{km} \\
-\ 0.\ 0\ 2\ 5\ 2\ 0\ 0\ 4\text{km} \\
\hline
5.\ 3\ 6\ 2\ 2\ 9\ 9\ 6\text{km}
\end{array}
$$

Problem 4. Simplify:

	2wk	6dy	7hr	8min	9sec
−	5wk	3dy	6hr	7min	8sec

		7dy	24hr	60min	
	− 4wk	2dy	5hr	6min	60sec
−	5̶w̶k̶	3̶d̶y̶	6̶h̶r̶	7̶m̶i̶n̶	8sec
	2wk	6dy	7hr	8min	9sec
−	2wk	3dy	22hr	58min	59sec

This is a takeaway from what you owe situation. − 5wk 3dy 6hr 7min 8sec is larger than 2wk 6dy 7hr 8min 9sec. You must reverse the order of the two quantities. Ensure you're taking Like from Like, and that you're exchanging equals for equals.

Problem 5. Simplify:

```
  6 3. 1 1 3 2
− 6 4. 0 0 2 1
```

This is a takeaway from what you owe situation. − 6 4. 0 0 2 1 is a larger number than 6 3. 1 1 3 2 You must reverse the order of the two quantities.

```
            9   9  10
     3    10  10   1  10
−  6  4.  0   0   2   1
   6  3.  1   1   3   2
−  0.  8   8   8   9
```

You are following the subtraction pattern. Starting from the smallest size, you exchange 1×0.001 for 10×0.0001. You can now subtract the .0002 from the 10×0.0001 that you had from the exchange, giving you a remainder of 0.0008. You add this to the 0.0001 you had, giving you a remainder of 0.0009.

You are now working on the next larger size. You can't take 0.003 away from 0.001 You need more 0.001. You don't have any in the next size, you must get them from what you do have, the Ones. You exchange 1×1 for 10×0.1 but 0.1 isn't what you need, it's 0.001. You continue exchanging until you reach size 0.001. You finally exchange 1×0.01 for 10×0.001. You continue the process, of subtraction until you have subtracted the final size. You end up with a final remainder of: 0.8889.

Problem 6.

	7dy	24hr	60min	
6wk	3dy	2hr	1min	60sec
~~7wk~~	~~4dy~~	~~3hr~~	~~2min~~	3sec
− 5wk	6dy	4hr	6min	8sec
1wk	4dy	22hr	55min	55sec

After each step is taken, ask yourself, why that Concept or Rule was used.

Problem 7.

```
       10         10  10         10        10 10 10 10 10 10
 - 8   4  10  2   4   1  10  6   8  10  3   1  2  0  6  2  3  10
 - 9   5   2  3   5   2   4  7   9   3  4   2  3  1  7  3  4   3
   7   7   9  2   8   6   7  6   9   5  3   7  6  2  8  9  5   4
 - 1   7   3  0   6   5   7  0   9   8  0   4  6  8  8  3  8   9
```

Notice in the original problem it was a "take away from what you owe" situation.

Mental tools are what you need and use in thinking and analysis. The number of ideas (mental tools) you use and logically put together in a problem-solving situation is how we measure the level of HIGHER ORDER THINKING (Vygotsky).

Solving a problem using a rote memorized problem-solving procedure, will not increase your ability to think outside the square, or increase your problem-solving ability. You only become more proficient solving the problems that can be solved using your rote learned procedure.

With the problem below, solve, by verifying with yourself, that you are confident you have used the appropriate Concept, Rule and Why, for each step of the solution process. When in doubt use problems 1 and 2 to confirm that your thinking and reasoning is correct. Each step you've done must be backed up by either a Concept or Rule . This ensures what you've done makes sense and verifies the result. It also helps build confidence. A What, How, and Why solution, is given on the next page.

Problem 8.

$$
\begin{array}{r}
9\ 1\ 4\ 5\ 3\ 2\ 4 \\
-\ 8\ 2\ 5\ 7\ 4\ 9\ 6 \\
\end{array}
$$

$$
\begin{array}{ccccccc}
 & 10 & 10 & 10 & 10 & 10 & \\
8 & 0 & 3 & 4 & 2 & 1 & 10 \\
9 & 1 & 4 & 5 & 3 & 2 & 4 \\
-\ 8 & 2 & 5 & 7 & 4 & 9 & 6 \\
\hline
 & 8 & 8 & 7 & 8 & 2 & 8 \\
\end{array}
$$

It becomes confusing if you put a zero at the very beginning of the number and wouldn't make sense.

CHAPTER 6

Understanding Multiplication, Algebraic Variables and the "Three One Rules"

In multiplication the incongruence of using English to explain the mathematics symbols and language becomes more of an issue. For example, in English the meaning of the Multiplication symbol "×" is communicated through several words, Multiply, Times, Product, Multiple Addition, etc. In Mathematics a symbol, word, or sentence, has only one meaning. It's this power that enables the language to be universal, and to communicate, Clearly, Correctly, and Concisely (The three Cs).

It's been mentioned previously that Mathematics' problem-solving power is using Concepts, Rules, and Why (Reflection, Visualization of patterns):

- The Concept is the shortest string of information that carries the understanding, the "What to do" (Vygotsky).
- The Rule is the shortest string of information that carries the "How to do it".

Mathematical Thinking is all about learning to use mental tools, Concepts, Rules, and Why. This type of learning is very important. The number of Concepts and Rules that you use during a problem-solving event is a measure of your ability to think at a Higher Level (Vygotsky). Mathematics is a problem-solving, logical, scientific language, which is unique as a language, as well as a subject, which has great power for increasing one's ability to think, reason, and apply, scientific thought.

Multiplication, when taught using Concepts and Rules, helps develop student's ability to concentrate for a longer period, and develop higher level thinking. For these reasons new learning should favour the ability to use Concepts and Rules.

In Mathematics the meaning of the Multiplication symbol "×" has only one meaning, it means, "Groups of".

Multiplication Concept - The multiplication symbol "×" means "groups of".

Multiplication Rule – Repetitive addition. You are adding identical value groups together, a given number of times.

Four groups of three means: $4 \times 3 = 3 + 3 + 3 + 3$
Eight groups of seven means: $8 \times 7 = 7 + 7 + 7 + 7 + 7 + 7 + 7 + 7$
The value of four groups of three means: $3 + 3 + 3 + 3 = 12$
The value of eight groups of seven means: $7 + 7 + 7 + 7 + 7 + 7 + 7 = 56$

You'll notice as you work through the Chapters that the same idea may be revisited a number of times, usually from a slightly different point of view. This repetition is done deliberately. We are working with important ideas, it's important to refresh what you have learned again, to reinforce its importance and retention.

To better understand multiplication the first Mental Tool the student should endeavour to use and learn is the "Two Digit Grouping Table" (commonly referred to as The Times Table)

Two Digit Grouping Table - The values of all the possible groupings of two single digit Numerals.

Unfortunately, when the meaning of the Two Digit Grouping Table was translated into English it was called the Times Table. This is, confusing, and lacks mathematical understanding. When first presented to beginning students, it might be thought of as a, "Table of the time it takes to do things", or it's about time. To be appreciated, Mathematics must always make sense and be relevant. If we are to understand Multiplication, we must first realize what the possible values any two-digit numbers can have when multiplied together. To help visualize these values we make a table.

Two Single Digit Grouping Table. Please note, in the font used in this book there is very little difference in appearance between the multiplication symbol and the alphabet letter x.

×	1	2	3	4	5	6	7	8	9	10
1	1	2	3	4	5	6	7	8	9	10
2	2	4	6	8	10	12	14	16	18	20
3	3	6	9	12	15	18	21	24	27	30
4	4	8	12	16	20	24	28	32	36	40
5	5	10	15	20	25	30	35	40	45	50
6	6	12	18	24	30	36	42	48	54	60
7	7	14	21	28	35	42	49	56	63	70
8	8	16	24	32	30	48	56	64	72	80
9	9	18	27	36	45	54	63	72	81	90
10	10	20	30	40	50	60	70	80	90	100

We are always trying to learn something new from each new idea. Let's look at the Single Digit Multiplication Table more scientifically. Scientifically means working with patterns. Yes, we notice we are doing a lot of multiple addition. Wait, look at the Ten's column. That's a repeat of the One's column only ten groups larger.

Eureka! We can use what we learned in the Number System Chapter, that ×10 means move the decimal point. That's exactly what the last column shows. That means we only need to know the two-digit grouping table up to 9 × 9 because grouping with 10 simply means move the decimal point. The true value of learning the Single Digit Grouping Table will become more visible. As we progress through the chapter, we'll understand how important and valuable the two-digit grouping table is.

In all the examples below $10^0 = 1$. We will investigate and confirm this in a later Chapter.

8×10^0 (8 groups of 10^0) is a discrete number. You have eight single units of size 1. This means you have eight groups of One.

8×10^1 is a discrete number. You have eight single units of size 10. You have 8×10 which is eighty. You've just added a zero to the 8.

28×10^1 is a discrete number. You have twenty-eight single units of 10. You have 28×10. Simplified, you have two hundred and eighty. You have simply added a zero to the 28.

Continuing this line of thought, the Number System demonstrated that:

26 is identical in value to - $2 \times 10^1 + 6 \times 10^0$
276 is identical in value to - $2 \times 10^2 + 7 \times 10^1 + 6 \times 10^0$
5,739 is identical in value to - $5 \times 10^3 + 7 \times 10^2 + 3 \times 10^1 + 9 \times 10^0$

We now can clearly see that every number can be displayed as a single digit with a grouping of 10.

So far, we only know how to multiply single digits. To be able to multiply more complex number arrangements we will need different types of grouping symbols to help display what we are multiplying. We use parenthesis ( ), brackets [ ], and Braces { }. They mean everything between the pair of symbols is treated as a single unit and acted on accordingly. We will use this knowledge and understanding when we get into multiplication. For example, we may have to multiply several elements by the same value at the same time. We'll demonstrate how this is done below.

3(2+7) means multiply everything in the parenthesis by 3. What's in the parenthesis? (2+7). We can either multiply each element by 3 and add the products together, or do the addition first then multiply by 3. Each method will give the same answer

3(2+7) = 3(9) =27 or 3(2+7) = 3×2 + 3×7= 6 + 21 = 27 Either way is following the Rules! Thus, both answers are identical.

Let's try 3(2x + 3y + 2A). Where x, y, and A are real numbers and 2x = x+x, 3y = y+y+y, and 2A = A+A

3(2x+3y+2A) = 3×2x + 3×3y + 3×2A = 6x+9y+6A..

As is demonstrated above, if you're following the Concepts and Rules correctly, you'll get the correct answer. This practice, doing it a different way, while still following the Concept and Rule, is excellent confidence building practice as well as increasing your ability to put ideas together in an organised, logical manner.

The parenthesis ( ) is the first order symbol pair. To make sense, and to be clear, the first order operation is done first. The brackets

[ ] and the braces { } are second and third order procedures. Following are the procedures.

$$3[2 +4(2+1)] \implies 3[2 + 4\times3] \implies 3[2 + 12] \implies 3[14] = 42$$

For clarity and understanding I have shown all the steps. With more confidence and experience, you would do most of the steps mentally. Notice you did the first order procedure first and continued to do the second order operation. This procedure should be clear and understood. The next example is a three-order situation. I will put in most of the steps.

2{3 + 2[3+2(3 + 4)]} In mathematics everything is clear, logical, understandable, and confidence building. You must follow the order of operation exactly!

$$2\{3 + 2[3+2(3 + 4]\} \implies 2\{3 + 2[3+2(7)]\} \implies 2\{3 + 2[17]\} \implies 2\{3+34\} \implies 2\{37\} = 74$$

6{3+5[3+2(6−3)]} = 6{3+5[3+2(3)]} = 6{3+5[3+6)]} = 6{3+5[9)]} = 6{3+45]} = 6{48} Universally this answer would be obtained. In many problem-solving situations the digits would include algebraic symbols. Let's do two more. You can fill in the steps.

27{15−13[25−3(20−3×4)]} =27{15−13[25−3×8]}=27{15−13[25− 24} = 27{15−13} = 54

4{3+6[58−3(24−18)]}=4{3+6[58−3(6)]}=4{3+6[40]}= 4{243]}= 972

Following are six problems with their answers for the purpose of practice, memory recall, perseverance, patience, and confidence building. Use the calculator sparingly to help. You are practicing and learning to put many simple ideas together in an organized logical manner. That's a very important life skill. It's important that you are gaining the understanding and confidence while getting the correct answer. If your answer was not correct, no problem, go more slowly, check your working for each step, try and find where you went off track, work to understand. A lot of effort was required to get the correct answer. Don't get frustrated. You can do it! You've had to put a lot of new ideas together to get the answer. The more time you spend on these 6 problems, the more you'll benefit from it. When you finally get the correct answer, well done! You're becoming a good, confident, conscientious thinker and analyser.

1. $23\{31 + 8[63 + 27(45 + 92)]\}$ Ans: 692,921

2. $-13\{8,321 + 8[47\ 19(54 - 92)]\}$ Ans: −37,973

3. $63\{13 + 6[39 - 13(133 - 81)]\}$ Ans: −239,967

4. $7\{-4 + 9[3 - 7(-17 + 24]\}$ Ans: −2926

5. $53x + 5[2x + 4(3x + 1)]$ Ans: 123x + 100

6. $3x\{2x + 3[3x + 2(3 + x)]\}$ Ans: 54x + 33x²
        Hint $x \times 11x = 11x^2$

The above process is problem-solving, requires deeper levels of scientific, logical, thinking, and following Concepts and Rules which complements how the brain works. The ability to be able to think and analyse to make better decisions is so very important. It makes more sense to teach Mathematics in a way that will help students reach higher levels of thinking, preparing them for later life problem-solving.

With our deeper levels of thinking and the Concepts and Rules we now have, let's revise what we've learned.

You are very familiar working with the pattern of single digits. Can we use this experience when the multiplication involves larger numbers? Let's see if our present knowledge can be applied to see if we can use the single digit Table to multiply larger numbers.

For example: Let's investigate 27 × 18 a multiplication problem involving two digit multiplied by two digit. The multiplication process is made up of two parts:

7×18 + 20×18

simplifying 7×18 we have:	$7 \times (1 \times 10^1 + 8 \times 10^0)$	
simplifying further we get	7×10 + 7×8	7 0
resulting in	70 + 56	5 6

simplifying 20×18 we have:	$2 \times 10^1 \times (1 \times 10^1 + 8 \times 10^0)$	
simplifying further we get	$2 \times 10^1 \times 10^1 + 2 \times 8 \times 10^1$	
	$2 \times 10^2 + 2 \times 8 \times 10^1$	1 6 0
resulting in	200 + 160	2 0 0
		1
	Total	4 8 6

We started with two single-digit multiplication but now realize we can arrange the numbers so that we only have to multiply single digits. The multiplying of the groups of Ten show where the decimal point ends up from the multiplication process.

We can arrange the above
so that it appears more like
the standard way of doing
Multiplication

$1 8 = 1 \times 10^1 + 8 \times 10^0 \times$
$\times 2 7 = 2 \times 10^1 + 7 \times 10^0$
$= 5 6 = 7 \times 10^0 \times 8 \times 10^0 = 56 \times 10^0$ All done in
$7 0 = 7 \times 10^0 \times 1 \times 10^1 = 7 \times 10^1$ head using
$1 6 0 = 2 \times 10^1 \times 8 \times 10^0 = 16 \times 10^1$ the procedure
$2 0 0 = 2 \times 10^1 \times 1 \times 10^1 = 2 \times 10^2$ carry
1
Total 4 8 6

Simplify: 6{3+5[3+2(6−3)]}

6{3+5[3+2(6−3)]} = 6{3+5[3+2(3)]} = 6{3+5[3+6)]} = 6{3+5[9)]} = 6{3+45]} = 6{48} Universally this answer would be obtained.

In many problem-solving situations the digits would include algebraic symbols. Let's do two more. You can fill in the steps.

$27\{15-13[25-3(20-3\times4)]\} = 27\{15-13[25-3\times8]\} = 27\{15-13[25-24\} = 27\{15-13\} = 54$

$4\{3+6[58-3(24-18)]\} = 4\{3+6[58-3(6)]\} = 4\{3+6[40]\} = 4\{243]\} = 972$

Following are six problems with their answers for the purpose of practice, memory recall, perseverance, patience, and confidence building. Use the calculator sparingly to help. You are practicing and learning to put many simple ideas together in an organized logical manner. That's a very important life's skill. It's important that you are gaining the understanding and confidence while getting the correct answer. If your answer was not correct, no problem, go more slowly, check your working for each step, try and find where you went off track, work to understand. A lot of effort was required to get the correct answer. Don't get frustrated. You can do it! You've had to put a lot of new ideas together to get the answer. The more time you spend on these 6 problems, the more you'll benefit from it. When you finally get the correct answer, well done! You're becoming a good, confident, conscientious thinker, and analyser.

I realize that you already have simple easy to follow procedures for doing multiplication problems which are much less laborious than what we are presently doing with Concepts and Rules. We are only using Mathematics as a vehicle for enhancing and developing your ability to; problem-solve, think and reason at a higher level, confidently use Rules and Concepts. These skills will last a lifetime and the ideas you will be learning will be transferable to your everyday life problem-solving experiences. Your rote-learned simple procedures for solving mathematics problems will simply fade with time, unless you are frequently using them. Your ability to think and reason will last a lifetime.

In Mathematics we have the symbol × which means groups of. That symbol facilitates the learning of the Two Single Digit Grouping Table and the number system. 10×1 becomes 1×10, 10×10 becomes 1×100, 10×100 becomes 1×1000 and so on. We can see another pattern "×10 " is an instruction to move the decimal point. This leads to an improvement of the pattern. Each improvement increased understanding.

From time to time throughout this book you may read something on which we have touched before. This is merely to reinforce information you have previously learned to help you remember.

INDEX OF 10 THE SIZE
$$1 = 10^0 \text{ ONES}$$
$$10 = 10^1 \text{ TENS}$$
$$100 = 10^2 \text{ HUNDREDS}$$
$$1000 = 10^3 \text{ THOUSANDS} \quad \text{This pattern continues indefinitely.}$$

The pattern above demonstrates the **Power of Ten Rule. ×10** means, move the decimal point. The index of 10 tells you how many places to move, and the sign of the index tells you the direction to move + to the right and minus to the left.

For example, $6×1000 = 6×10^3 = 6×10×10×10 = 6,000$. Recognizing patterns is very important. When you are recognizing, using or modifying patterns you are thinking scientifically. Only human beings have this ability.

Many students are not being taught how to think logically. Mathematics is a problem-solving, logical, scientific language which has great value for improving ability and confidence in problem-solving as well as increasing ability of higher levels of thinking. The learning may be difficult at first but persevere. You will gain much from the experience.

Let's, mathematically investigate 234 × 567. The main reason we will be working with this problem is to gain experience working with Concepts and Rules which are transferable for working in other problems. It's completely different from learning a rote learned procedure to pass an assessment which gives little Mathematics understanding. In that situation the knowledge learned was very specific and little understanding could be carried over to more complicated problems.

234 means, and is the shorthand method for writing, 2×100 + 3×10 + 4×1
567 means, and is the shorthand method for writing, 5×100 + 6×10 + 7×1

$$234 \quad 2 \times 100 + 3 \times 10 + 4 \times 1$$
$$\times\ \underline{567} \quad 5 \times 100 + 6 \times 10 + 7 \times 1$$

Our effort now goes towards reducing the multiplicative process to single digit multiplication and index notation

The index 10 notation for this problem will include 10^5 10^4 10^3 10^2 10^1 10^0 It makes sense to break the problem down into four main parts.

7×234   60×234   500×234 $234 = 4 \times 10^0 + 3 \times 10^1 + 2 \times 10^2$

$7 \times 4 \times 10^0 = 28 \times 10^0 = 2 \times 10^1 \times 10^0 + 8 \times 10^0 = 2 \times 10^1 + 8 \times 10^0$
$7 \times 3 \times 10^1 = 21 \times 10^1 = 2 \times 10^1 \times 10^1 + 1 \times 10^1 = 2 \times 10^2 + 1 \times 10^1$
$7 \times 2 \times 10^2 = 14 \times 10^2 = 10 \times 10^2 + 4 \times 10^2 = 1 \times 10^3 + 4 \times 10^2$

$6 \times 10^1 \times 4 \times 10^0 = 24 \times 10^1 = 2 \times 10^1 \times 10^1 + 4 \times 10^1 = 2 \times 10^2 + 4 \times 10^1$
$6 \times 10^1 \times 3 \times 10^1 = 18 \times 10^2 = 1 \times 10^1 \times 10^2 + 8 \times 10^2 = 1 \times 10^3 + 8 \times 10^2$
$6 \times 10^1 \times 2 \times 10^2 = 12 \times 10^3 = 1 \times 10^1 \times 10^3 + 2 \times 10^3 = 1 \times 10^4 + 2 \times 10^3$

$5 \times 10^2 \times 4 \times 10^0 = 20 \times 10^2 = 2 \times 10^1 \times 10^2 = 2 \times 10^3$
$5 \times 10^2 \times 3 \times 10^1 = 15 \times 10^3 = 1 \times 10 \times 10^3 + 5 \times 10^3 = 1 \times 10^4 + 5 \times 10^3$
$5 \times 10^2 \times 2 \times 10^2 = 10 \times 10^4 = 1 \times 10^5$

The next step is simple: Just add up all the groups of Like things.

All the above took a superhuman effort. By following all of the Concepts and Rules required (there were only a few different ones but a lot of repetition), you would be able to multiply any two numbers together large, small or of a different Base size.

The practice you've had doing this will give you the patience, tenacity, understanding, and confidence, to work on any difficult situation. Many problems took years to solve, and there are many problems we're still working on.

Each step we take should be getting us closer to achieving a pair of single digits with their individual index of Ten. If we can achieve that, we can use the Single Digit Grouping Table knowledge (Times Table) and your understanding of the number system. Our result will be a number with sizes increasing as we go from right to left. A table will help us organize our results in a logical manner. We will use the first column on the right for the size ones, the second for size tens, third for hundreds and so on. This next time it will be much easier as we'll be taking advantage of the many skills we gained from the earlier problems.

We will start by calculating the value of 7×234, one step at a time.

First, seven groups of four ones - $7 \times 4 \times 1$ $= 28 \times 1$,
Second seven groups of three tens - $7 \times 3 \times 10$ $= 21 \times 10$
Third seven groups of two hundred - $7 \times 2 \times 100$ $= 14 \times 100$.

Our number system is made up of groups of ten. Every time we have a group of ten it is packaged up and incorporated into the next larger size, thus:

$7 \times 4 \times 1 = 28 \times 10^0$ and is identical in value to $2 \times 10^1 + 8 \times 10^0$
$7 \times 3 \times 10 = 21 \times 10^1$ and is identical in value to $2 \times 10^2 + 1 \times 10^1 = 2 \times 10^2$ $+ 1 \times 10^1$

$7 \times 2 \times 100 = 14 \times 10^2$ and is identical in value to $10 \times 10^2 + 4 \times 10^2 = 1 \times 10^3 + 4 \times 10^2$

Our total so far from the 7 × 234 (seven groups of 234) process would be:

HUNDRED THOUSAND 10^4 ×10,000	TEN THOUSANDS 10^3 ×1000	THOUSAND 10^2 ×100	HUNDREDS 10^1 ×10	TENS 10^0 ×1	ONES
				2	8
			2	1	0
		1	4	0	0

Notice, that we always insert a zero when we don't have any of that size.

We continue by calculating the value of 6 ×10 × 234 one step at a time.

sixty groups of $4 \times 10^0 = 6 \times 10^1 \times 4 \times 10^0 = 24 \times 10^1 = 2 \times 10^2$
sixty groups of $3 \times 10^1 = 6 \times 10^1 \times 3 \times 10^1 = 18 \times 10^2 = 10 \times 10^2 + 8 \times 10^2$
sixty groups of $2 \times 10^2 = 6 \times 10^1 \times 2 \times 10^2 = 12 \times 10^3 = 10 \times 10^3 + 2 \times 10^3$

Our subtotal including the of 6 ×10¹ × 234 multiplication process is:

HUNDRED THOUSANDS 10^5 ×100,000	TEN THOUSANDS 10^4 ×10,000	THOUSAND 10^3 ×1000	HUNDREDS 10^2 ×100	TENS 10^1 ×10	ONES 10^0 ×1
				2	8
			2	1	
		1	4	0	0
			2	4	0
		1	8	0	0
	1	2	0	0	0

We continue by calculating the value of 5×100 × 234 one step at a time.

First, five hundred groups of four ones - 5 ×100 × 4 × 1 = 20 × 100
Second. five hundred groups of thirty - 5 ×100 × 3 ×10 = 15 ×1000
Third, five hundred groups of two hundred - 5 ×100 × 2 ×100 = 10 ×10,000

Our final subtotal including of 5×100 × 234 multiplication process is:

HUNDRED THOUSANDS 10^5 ×100,000	TEN THOUSANDS 10^4 ×10,000	THOUSAND 10^3 ×1000	HUNDREDS 10^2 ×100	TENS 10^1 ×10	ONES 10^0 ×1
				2	8
			2	1	
		1	4	0	0
			2	4	0
		1	8	0	0
	1	2	0	0	0
		2	0	0	0
	1	5	0	0	0
1	0	0	0	0	0
1	3	2	6	7	8

When we total what we have, we follow the same pattern. The maximum that you can have in any one size is 9 packages, when we have more than 9, we package the excess into groups of ten and include that amount to the total of the next size column. For example, when we added the number of hundreds the total was 16. The 6 stayed as the number of hundreds and the package of 10 Hundreds was mentally included when we added the thousands (11 + 1 = 12). This pattern continued for all of the other totals.

Analysing the above you can see several repeating patterns. Problem-solvers are always looking for patterns to make things easier. Calculating the above grouping (multiplication) problem by using your mental tools builds problem-solving ability as well as confidence in your ability to put several different ideas together to discover new ideas. I will leave it up to you to figure out why the multiplication procedure that you have been taught in arithmetic lessons always works. It will be a very good thinking exercise.

Below is a problem for you to practice on. Remember you are learning how to put ideas together in an organized, logical manner. It's an important skill which will be very important for you when it comes to life long problem-solving! Please follow the pattern that we used for 27 × 18. You can set your own problems to solve as you are now in control of your own learning. You can make the problem as challenging as you like. You can easily check your answer with a calculator AFTER YOU HAVE SOLVED IT YOURSELF! Below is one I want you to do as a starter.

36 × 25 Here is the heading for your start. 10^3 10^2 10^1 10^0

The Base Ten Number system is based on groups of ten. Every time you get a group of ten, it becomes one of the next larger sizes. To consolidate your understanding of what you've done with Base 10 multiplication, we'll have a go at doing multiplication in a different Base Number. This will be a big challenge for you, translating your understanding and wisdom to a different situation.

I'll demonstrate on two problems, then it's up to you to try one yourself. It will be a big challenge, but you will gain a lot of understanding and confidence in yourself as a logical problem-solver from trying it. We'll multiply two Base eight numbers. You can always check your answer by translating both numbers into a Base 10 number after you finish.

What you first must realize when you are working with other Bases is that you need to know the pattern of the Base Number. The pattern for Base 10 is below:

$$\ldots 10^4 \quad 10^3 \quad 10^2 \; 10^1 \; 10^0 \; 10^{-1} \; 10^{-2} \quad 10^{-3} \quad 10^{-4} \ldots$$
$$\Longleftarrow 10{,}000 \; 1{,}000 \; 100 \; 10 \; 1 \; 0.1 \; 0.01 \; 0.001 \; 0.0001 \Longrightarrow$$

For Base 8 the pattern would be:

$$\Longleftarrow \ldots 8^4 \quad 8^3 \quad 8^2 \; 8^1 \; 8^0 \; 8^{-1} \quad 8^{-2} \quad 8^{-3} \quad 8^{-3} \ldots \Longrightarrow$$
$$4{,}096 \quad 512 \quad 64 \; 8 \; 1 \; 1/8 \; 1/64 \; 1/512 \; 1/4096$$

Given: 7654_8 determine its Base 10 value.

7654_8 represents $4\times8^0 + 5\times8^1 + 6\times8^2 + 7\times8^3$ the index of 8 tells you how many times to multiply the 8 by itself: Converting 4567_8 to its Base. 10 value we have

$$4\times8^0 = 4\times1 = 4$$
$$5\times8^1 = 5\times8 = 40$$
$$6\times8^2 = 6\times8\times8 = 6\times64 = 384$$
$$7\times8^3 = 7\times8\times8\times8 = 7\times512 = 3584$$

This will be a big challenge. It'll be a big accomplishment for you trying something completely new. I'm sure if the challenge on the basketball court was trying to get the basketball into the hoop, you'd give it a go, and if you missed , you'd modify what you've just done and try again. Well, you must do the same thing with intellectual problems. You'll accomplish a lot more by trying again. The effort in having a go solving the problem $35_8 \times 42_8$ above, requires your complete attention, concentration, and effort. You're concentrating on using your understanding of the Base 10 Number system and following the same pattern with the new sizes. For referral, solution is on the next page.

35_8 means and is shorthand for $3\times8^1 + 5\times8^0$ means $3\times8^1+5\times8^0=29$ Base 10

42_8 means and is shorthand for $4×8^1 + 4×8^0$ identical in value to 36 Base 10

	4,096	512	64	8	1
For Base 8 the pattern would be.	8^4	8^3	8^2	8^1	8^0

Given. $35_8 × 42_8$ Your result Table

Base 10 Equivalent	4096	512	64	8	1
	×8×8×8×8	×8×8×8	×8×8	×8	×1
This is your heading for Base 8	8^4	8^3	8^2	8^1	8^0

Given. $35_8 × 42_8$ Things to remember working on the problem. You only have nine digits to work with 0 - 8.

$5×8^0×2×8^0 = 10×8^0 = (8+2)×8^0 = 8^1×8^0+2×8^0 = 8^1 + 2$

$5×2×8^0=10×8^0=1×8^1×8^0+2×8^0 = 1×8^1 +2×8^0$	1 2
$5×4×8^1=20×8^1=2×8^1×8^1+4×8^1 = 2×8^2 +4×8^1$	2 4 0
$3×8^1×2×8^0=6×8^1 =$	6 0
$3×8^1×4×8^1=12×8^2=(4 + 8^1)×8^2 = 4×8^2 + 8^3$	1 4 0 0
	1
TOTAL	1 7 3 2

Checking our Answer- $1732_8 = 512 + 7×64 + 3×8 + 2 = 986$

$35_8 × 42_8 = (24+5)(32+2)=29×34=986$

You are now able to make your own problems and check your answer by converting them to Base 10 like was done in the demonstration.

Let's look at a familiar different simple grouping knowledge problem again. This time looking at it from a different point of view to see if we could discover something new.

In Mathematics we communicate with concrete and abstract symbols. Let's explore, with What, How, and Why, thinking, the topic Multiplication, to see where it leads. We'll use our Symbol for multiplication. × = groups of . Remember, we are always looking for patterns, that means looking at things from a different perspective. Starting with something simple we have:

3 × 4 = 4 + 4 + 4 We know from our understanding of the Number System that 3 represents three ones, or a package of three. We could communicate 3 × 4 with concrete symbols:

3 × 4 = ○●●○ + ○●●○ + ○●●○

We could arrange these in any order without changing the value of what we have. Let's try the following pattern:

If we look at the pattern below from left to right, we have:

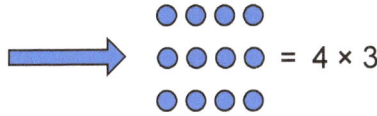

$$\longrightarrow \quad \begin{matrix} ○●●○ \\ ○●●○ \\ ○●●○ \end{matrix} = 4 \times 3$$

But, if we observe this scientifically, and look at the same Pattern and number of things, by looking top to bottom, we get a different perspective. Looking at the pattern from top to bottom or bottom to top doesn't change the value of what we have.

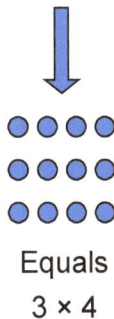

$$\downarrow$$

$$\begin{matrix} ○●●○ \\ ○●●○ \\ ○●●○ \end{matrix}$$

Equals

3 × 4

We see 3 × 4. If you were working in a shop or purchasing things you would be very aware of the number of things in a package. Four packages of three are identical in value to three packages of four. You would only purchase the package size you wanted.

Any quantity could be arranged in this grid type pattern. We have discovered a new idea. Any multiplication process can be expressed as a right to left grouping, or a left to right grouping. We can thus formalize the idea using algebraic symbols:

For any ab, ab = ba where a and b are real values

This new idea also means that you only need to memorize a little bit more than half the single digit grouping table. Each grouping can be expressed two ways, left to right or right to left. The brain works very efficiently calculating left to right or right to left. Which means, when learning the Times Table, you can decide which way you want to learn it. If 3×6 is easier memorize than 6×3 you could apply this pattern to the rest of the Tables and only memorize a little more than half of them. When you see 6×3 the brain will automatically say $3 \times 6 = 18$.

If the brain can read things right to left rather than the conventual left to right then it makes sense to do this. Will it try to make a new understanding of a statement if you deliberately read the idea right to left, instead of left to right? Let's look at the following:

Imagine you get your first holiday job at the supermarket. Because it's your first job you'll start off with a simple procedure. Your job is to answer the telephone. All the calls for lunch time apples orders will be routed to you. You'll put the number of apples required for the order in a bag for a lunch time pick-up. That's seems simple enough. The telephone orders you received on your first day are listed below:

3 apples + 2 apples + 2 apples + 3 apples +1 apple + 3 apples + 3 apples + 1 apple + 2 apples + 1 apples.

It's now lunch time so this part of your job ends. How many apples did you sell? 21

You're a conscientious, thinking person, and you realize it takes time to write out apples each time for each order. If the phone line is tied up because you're writing out an order your losing customers.

You think, how can you record the orders faster? Algebra! When you use an algebraic symbol to communicate it's able to communicate a symbol, a word, or a whole paragraph accurately and faster. The only stipulation is that the symbol must be defined, so that it's clear, concise, and makes sense.

You think and reflect. The first letter of what you are wanting to communicate is usually the start of the sound of what you're thinking of. Eureka! That's a good idea, and a new pattern to remember.

It makes sense to define A for apple. This is usually done by "Let A = Apple" so until you change the meaning of the symbol, A stands for apple. The next day you're ready with your new idea. Let A = apple.

Internationally, it's agreed when multiplying in algebra, you don't use the multiplication symbol between the numeral (quantity) and the algebraic symbol, or between algebraic symbols. For example, if a =2, b= 3 and c = 4 then:

"abc" means $2 \times 3 \times 4 = 24$ and $2abc = 2 \times 2 \times 3 \times 4 = 48$. Algebra can communicate sophisticated information and processes quickly and easily.

There are only four basic operations in maths:

- addition,
- subtraction,
- multiplication, and
- division.

If no operation symbol is shown it means multiply. For example: 5A represent five apples, the symbol A by itself would represents one apple, and 7A + 3A + 3A would represent the 13 apples that were sold.

Below is your sales record for the second day:

3A + A + 4A + 3A + 2A + 5A+ A + 4A + 2A + 4A + 3A + 2A + 4A + 2A + A

The phone wasn't tied up all the time, you had more phone orders, and when customers observed you weren't busy on the phone., you had cash purchases as well. How many apples did you sell this time? 41

This is only the second day at work, and you've almost doubled the sales. The manager's very impressed. You take advantage of his enthusiasm and say, "I could handle bananas as well". Do you think the boss would agree". Of course he would, you've already demonstrated you've got initiative, and your new idea worked. What symbol would you use for banana? You're learning to follow patterns so you would add to the list "B = banana".

This is your first order of the day you write 6A + 7B. What did you sell?

If you said six apples and seven bananas, you're already demonstrating you can see the advantage of algebra. You're still thinking and want to further impress the manager, so at the end of another successful day you tell him. "Some people like red apples, others like yellow or green ones, if we had more varieties we could sell more, and I could handle that". Again, the manager is impressed and could see you've got ideas and enthusiasm that are good for the business, so he agrees.

What symbols would you use for red, yellow and green? You've already started to use a pattern that's simple and makes sense, you'd use: R for red, Y for yellow, and G for green. Your new list is getting longer, so you put everything in alphabetical order for convenience and efficiency. Your list should look like the list below:

Let A = apple
    B = banana
    F = fruit
    G = green
    R = red
    Y = yellow

We have a new problem. How do we communicate the different colours efficiently?

In Mathematics when we want to communicate more information about a variable, you communicate the information with a smaller font symbol after the variable and at its foot. For example: if we wanted to say five red apples, and seven green bananas we would write $5A_R + 7B_G$. If we wanted to communicate six green bananas and ten green apples, you'd write $6B_G + 10A_G$.

In a previous chapter it was stated that Mathematics was a "Problem-solving, logical, scientific language". English doesn't always communicate clearly, correctly, and accurately. It is not like Mathematics. It is not efficient , convenient, or clear when it is used to communicate Mathematically. In English a symbol can have several different meanings, a word can have eight or more meanings, and a sentence can have thousands of different possible combinations. This can create big problems when trying to communicate mathematical understanding.

When you were taught to read and write in English, it was drummed into you that you always read and write left to right. Your brain is a very efficient problem-solving, logical, scientific organ. It never sleeps and is problem-solving, making sense of things, twenty-four seven. When it sees a more logical way of saying things it will kick in and read right to left, going against all you have been taught, about reading left to right. Consider the following algebraic statement using the defined meanings of symbols that we have used in the previous page. Please softly read out the following algebraic statement:

$$5A_R + 7B_G + 6B_Y + 10A_G + 10A_Y$$

Did you read it left to right, did it make more sense for some of it right to left If you let your brain do the thinking instead of just memorizing things you'd say five red apples, not five apples that are red. Once you start realizing you can sometimes get a better idea reading right to left, you'll look out for these possibilities.

When you go to work the next day armed with your new ideas and symbols, the manager tells you - "The person who handles the flowers is sick today, I don't have anybody else who could handle the department. Do you think you could handle it?". You don't want to disappoint him , so you accept.

You'll be selling flowers today; you also notice that, as well as red and yellow, you have two new colours white flowers and purple ones. What symbols would you use to represent flowers, white, and purple? You've already used capital F for fruit. Is there another F symbol that can keep the sound, stay with the same sort of articles and follow your new pattern? You think of lower-case f for flowers, W for white, and P for purple. have to add these new symbols to the list which should now look like the one below:

Let A = apple
    B = banana
    f = flower

F = fruit
G = green
P= purple
R = red
W = white
Y = yellow

By using upper- and lower-case letters, you have increased the number of symbols you know and easily recognise from 26 to 52.

Here's todays first sale: $8f_W + 17f_R + 10f_Y + 5f_P$. What did you sell on your first sale? Follow the same pattern as before. A second logical answer would be 40 flowers.

Tomorrow you're back with your original fruit department. Below is what you sell on your first sale.

$6A_G + A_R + 3A_G + A_R + 10B_G + 5B_Y$ What did you sell on your first sale? Follow the same pattern as before. A second logical answer would be 26 fruits.

Your brain works logically. Even if you were taught to do it one way, if the brain could see it makes more sense to do it another way, if you're in tune with your brain, you would do it the way the brain says going against everything you were told. Have you thought of the following yet? I'm sure given $5 you would say "five dollars", not "dollars that are five"!

The next step is, can you read a whole sentence that's written in English from right to left or left to right. This time we are talking about ideas. The answer is in some cases is YES, and that leads us to the Three One Rules, probably the most useful rules in Mathematics problem-solving. By being able to read it from right to left you are getting a completely different view of the situation, and this can lead to all sorts of new ideas and ways of doing things.

As mentioned previously, intellectual thinking and problem-solving requires Concepts and Rules. Each Concept and Rule is an intellectual tool. This cannot be overstated. The more intellectual tools in your tool kit the greater variety of problems you're able to get involved with. The Three One Rules have already been mentioned, so let's formulate them now. Your imagination is a very powerful place to visit for modifying old patterns and creating new ones.

Let's imagine you're three years old again. Mum and Dad have given you a Teddy Bear for your birthday. You love your Teddy and go to sleep with it wrapped in your arms. One night you have a marvellous dream. In the dream your Teddy invites three more Teddies over to play. You have a fantastic time in this dream, but when you wake up, how many Teddies do you have? Even a three-year-old knows the answer, only one. No matter how much you would like more, if you only have one of something, you only have one of them. We can formulize that statement into a Rule.

Anything multiplied by 1,
remains the same in value.

We will consider the above Rule to be a pattern, as it applies to anything. It's made up of two parts. Anything multiplied by 1, remains the same in value. If we look at the two parts as two ideas and read each idea from right to left, we come up with a new idea.

Anything remains the same in value,
when you multiply it by 1

That creates two new possibilities:

1. If a problem involves multiplication and the multiplication is causing you difficulty, you can eliminate the multiplication if, in doing so, you are, in effect, only multiplying the whole problem by 1.

2. If a problem doesn't involve multiplication but would become easier if it did, you can insert the required multiplication providing you are, in effect, only multiplying the whole problem by 1.

The original Rule has now become two very useful, and powerful, new Rules.

Let us now consider an idea involving division. From our understanding of Numbers we know that every number is made up of Ones. At the beginning that was very clear; one one, two ones, three ones, etc. until we got to ten ones, and then things became more complex. The common thought of counting by ones flew out the window. Let us now revisit the situation. If we had a group of six buttons (ones) on the table; how many groups of one could we make? The answer should be obvious, but surprisingly for many school children they'd have to have a serious think. Fortunately, it's very easy to demonstrate same, with a diagram.

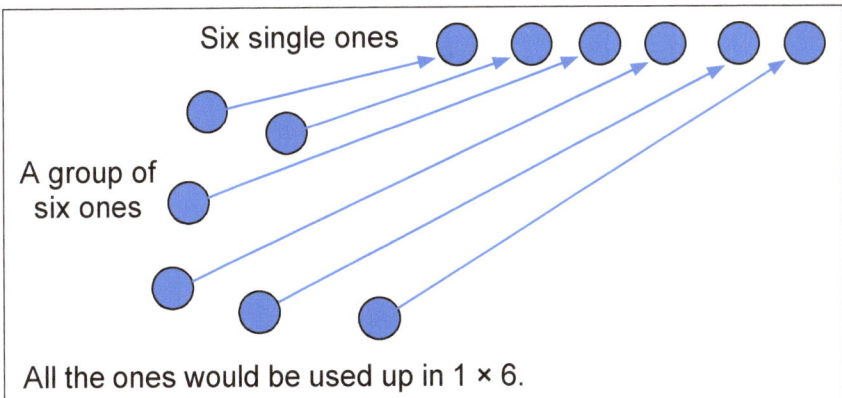

Six single ones

A group of
six ones

All the ones would be used up in 1 × 6.

Because every number is made up of ones we can formalize that idea to a Rule.

Any quantity divided by 1,
remains the same in value.

We will consider the above Rule to be a pattern, as it applies to anything. It's made up of two parts. Anything divided by 1 remains the same in value. If we look at the two parts, as two ideas and read each idea from right to left we come up with a new idea.

> Anything remains the same in value,
> when you divide it by 1

That creates two new possibilities:

3. If a problem involves division and the division is causing you difficulty, you can eliminate the division, if, in doing so, you are in effect only dividing the whole problem by 1.

4. If a problem doesn't involve division but would become easier if it did, you can insert the required division providing you are, in effect, only dividing the whole problem by 1.

As above, the original Rule has now become two very useful, and powerful new Rules.

The four new Rules which we have formulated, each involve the numeral one. That means we can integrate the two ideas; multiply by one and divide by one into the same problem, as multiplying by 1 or dividing by 1 won't change the value of the problem you are working on.

That gives us a fifth Rule.

5. We can use the symbol "× multiply or − divide" where − is the fraction line, $\frac{a}{b}$ meaning divide the a by the b where a and b are any number. We'll see much more about this fifth rule when we look at equations and fractions Chapter 7.

Obviously, there is something special about the 1 we are going to use, so let's put our effort into finding out more about getting the 1 that we need. We shall return to our group of six buttons. If we had six buttons on the table, how many groups of six can we make from six buttons? Well, the multiplication rule can help us here. If we only have one group of six, then we can only make one group of six from it. In other words, everything contains itself, and nothing, more, or less. We can formalize that idea into our third One Rule. If we are talking about division, we had better learn the mathematical terms:

Given A, B, and C, are any numbers, and the expression:

Anything divided by itself, is equal to 1. Excluding divide by zero

We can't divide by zero. In division as the denominator (bottom number) in a division process gets smaller and smaller the value of the division process gets bigger and bigger. There is no limit to how small the denominator gets, so there is no limit to how large the quotient (value) of the division can get. In other words, what's the largest number you can have? There's no limit. We use the infinity symbol (∞) in Mathematics to indicate the largest quantity or number,

The Three One Rules (They will be referred to as TOR and TOR1, 2, or 3)

TOR1: "Anything multiplied by One stays the same in value". This sentence is self-evident. If you only have one of something, you only have one of them. In Mathematics we strive to say things Clearly, Correctly, and Concisely. Looking at the idea from, left to right and right to left.

1 Anything multiplied by ONE, stays the same in value. The value remains the same if multiplied by ONE. **TOR1**

The difference in meaning is Subtle. The top line tells says you can multiply by ONE without changing the value.

The bottom line, if you want something to stay the same in value, you multiply by one.

2 Anything divided by ONE stays the same in value. The value stays the same if we divide by ONE. **TOR2**

Again, the difference in meaning is Subtle. When anything is divided by ONE it stays the same in value. The bottom line. If you want something to stay the same in value, you divide by ONE.

3 Anything divided by itself is 1. Exclude dividing by zero **TOR3**

The above rules were discovered by investigating the effect of reading the meaning or idea from right to left as well as from left to right. It's up to you to start investigating patterns, to see what happens when you read ideas from right to left as well as from left to right.

The power and efficiency provided by the Three One Rules will be demonstrated when working with fractions and equations in Chapter 8

To reinforce and give more practice, below is another example involving the grouping of large numbers.

We started multiplication by defining: "×" to means "groups of" and $4 \times 5 = 5 + 5 + 5 + 5$ and the value of $4 \times 5 = 20$

But what do we do when the numbers are larger? For example, 365×278. Whether we look at the expression left to right or right to left we are talking about a lot of groups. This is where the Times table comes in. You only need to know the combinations up to 9 × 9. When you are working with a "ten" you are working with a decimal point change. For example:

10 × 1 = 10	Again we see a pattern.	$10 = 10^1$
100 × 1 = 100	The index tells you how	$100 = 10^2$
1,000 × 1 = 1,000	many places to move the	$1,000 = 10^3$
10,000 × 1 = 10000	decimal place.	$10,000 = 10^4$

This idea was formalised by calling it the **The Power of Ten Rule.**

The sign of the index tells you the direction to move. Positive to right. Negative to the left. + to ⟶ right – to ⟵ left. If the index has no sign, it's Positive

× 10^4 (×**10**) means move the decimal point

The index four means move the decimal point to the right four places The index has no sign, so it is positive

What does the index value zero mean. We haven't figured out what 10^0 means. That's a job for the **Three One Rules.** Our power is the equation. Once we have an equation, we can do whatever we want to one side PROVIDING we do the equivalent to the other. We need an equation with 10^0 in it. No problem. If we want an equation with a particular value, set the value equal to itself.

For manipulating equations and fractions, the **Three One Rules** are the first set of tools to use. Our investigation should always describe the situation Clearly, Correctly, and Concisely. I know I said *when* we work with equations, but there's no harm in giving you the Equation Concept and Rule now.

Equation Concept – The left side of the equals sign has the identical value as the right side.

Equation Rule – You can do what ever you like to one side of the equals sign, providing you do the EQUIVALENT to the other side.

The Equation Concept and Rule provide the powerhouse for Mathematical investigations. Described another way: When are you in a situation that you have the power to do whatever you like? When you have your ideas in the form of an equation.

To investigate the meaning of x^0 ; Easy! Step (1) Just set x^0 equal to itself. We now have what we want on the left of the equals. If we just use our Concepts and Rules on the right-hand side without changing the value of the right hand side, we will have found out what the left side is equal to!

$$\text{Step 1} \qquad 10^0 = 10^0$$

We can multiply the right side of the equation by one, without having to do anything to the left side. We want the left side to stay the same in value. It will be the right side which will tell us what the left side ends up equal to!

We are working with an index notation element, so we will multiply by an index notation element. We want our conclusion to be robust so we will use an index of any value (x). We want the left side to remain itself. That means whatever we do to the right side in effect must be equal to ONE. The perfect tools for that are division and multiplication! TOR1 and TOR2. We are working with an element with an index of 0, so it makes sense for us to also work with elements which have indices. TOR2 anything divided by itself excluding dividing by zero. This always gives us ONE. This is step (2). That means we will be working with fractions. We need more intellectual tools, the **Fraction Concept** and **Rule**, and the **Multiplication of Fractions Rule.**

The Multiplication of Fractions Rule-when multiplying fractions: Numerators are multiplied together, Denominators are multiplied together, after first checking to see if any division can be done.

$$\text{Step 2} \quad 10^0 = 10^0 \times \frac{10^x}{10^x} \quad \text{TOR 1 and TOR 3}$$

At the present time 10^0 is not a fraction. Step 3. We will make it a fraction (A fraction is a division instruction that is, the Concept for a fraction. See following Chapter on Fractions) without changing its value by dividing it by 1

$$\text{Step 3} \quad 10^0 = \frac{10^0}{1} \times \frac{10^x}{10^x} \quad \text{TOR 2}$$

In Step 4, we will use The Rule for multiplying fractions, and the rule for multiplying same value quantities with indices.

The Rule for multiplying same value quantities with an index is to add the indices together. For example:

$2^2 \times 2^3 = \underline{2 \times 2} \times \underline{2 \times 2 \times 2} = 2^5$

Step 4

$$10^0 = \frac{10^0}{1} \times \frac{10^x}{10^x} = \frac{10^0}{1} \times \frac{10^x}{10^x} = \frac{10^{(x+0)}}{10^x} = \frac{10^x}{10^x} = 1$$

Our result is $10^0 = 1$ We correctly used our Concepts and Rules. Our result occurred when we used x for any value, our method was robust. The Equation Concept tells us that the left side of the equation has the identical value as the right side. We could have used any number instead of 10 and received the same result. We are then confident that for any number x:

$$x^0 = 1 \quad \text{Where } x \text{ represents any number}$$

Setting something you don't know, or understand, equal to itself so that you can manipulate the right-hand side using Concepts and Rules which are always correct, and without changing the value of the right hand side, is a very powerful mental tool to use for exploring and investigating new ideas.

The Power (Index) of Ten Rule

When we are working with the multiplication of a numeral multiplied by a 10^x where x is any value, the Positive Index tells you how many decimal places to move the decimal to the right, and a Negative Index tell you how many places to move the decimal to the left. Our Number System is built on this Rule Every time we have ten of any one size, it becomes one more of that size. For example: ten groups of I, become a ten, ten tens become a hundred and so on. Likewise, when you divide by ten it becomes one less. Less means minus direction. See examples below:

a) $4.34 \times 10^3 = 434$

b) $.005327 \times 10^4 = 53.2.$

c) $1234. \times 10^3 = 1234000$

d) $.063 \times 10^{-4} = 0.000006$

e) $5^3 \times 5^2 \times 5^{-3} = 5^2 = 5 \times 5 = 25$

f) $4^3 \times 4^{-2} \times 4^{-3} \times 4^3 = 4^2 = 16$

g) $2^2 \times 3^2 \times 2 \times 3^2 = 2^3 \times 3^4$

h) $(0.2)^2 = 0.2 \times 0.2 = 2 \times 10^{-1} \times 2 \times 10^{-1}$
$= 4 \times 10^{-2} = 0.04$

When you are not sure that your answer is correct, or just to increase confidence, if you have understanding, you can try another method of solution. If each time you follow the Concepts and Rules correctly, each answer will be the same and will be correct. I hope you now understand how important Concepts and Rules are in problem solving and increasing your intellect

You might now have another question. What does 4^{-3} mean? You have several mental tools to use for this problem. Using a previous pattern you remember your strength and power is the equation. Set the number equal to itself.

$4^{-3} = 4^{-3}$ You know what a positive index means. The first step would be to eliminate the negative index. You would multiply by 4 with a positive index. You have the power to do whatever you want. You don't want to change the left side so you make the multiplier equal to 1. Anything divided by itself is 1. The obvious choice would be $\frac{4^3}{4^3}$

This gives you:

$$4^{-3} = 4^{-3} \times \frac{4^3}{4^3} = \frac{4^{-3}}{1} \times \frac{4^3}{4^3} = \frac{4^{(-3+3)}}{4^3} = \frac{4^0}{4^3} = \frac{1}{4^3}$$

In the above investigation you used Concepts and Rules in an organized logical scientific way. You have come up with a pattern which is reproducible. You can Generalize your new Rule.

For any real number x, or y – $x^{-y} = \dfrac{1}{x^y}$

By solving problems this way, you are increasing your ability to think at a higher level as well as increasing your ability and confidence in problem-solving. You are using abstract symbols in problem-solving in a very efficient and effective way.

One more example for revision.

$45 \times 10^3 \times 10^3$ Following the pattern it means move the decimal to the right three places, then three again. This is a total of six to the right which is x 10^6. $45 \times 10^6 = 45000000$.

This last example leads us to the main rule for multiplication between the same elements (values) that have an index. You add the indices.

Index Rule: When you multiply identical quantities which have an index you add the indices.

For example: $4^5 \times 4^5 \times 4^5 \times 4^5 = 4^2$ $9(4+5)(4+5) = 9 \times 9 \times 9 = 9^3$

$(x^2y^3)\ (x^2y^3)\ (x^2y^3) = x^6y^9$

As you gain more experience and expertise following Concepts, Rules, and patterns, you will realize if you follow the Concepts, Rules, and patterns exactly, you have complete confidence that your answer is correct. That's why understanding is so important. Your confidence is based on how well you understand what you're doing. Sadly, most students have not realized that Mathematics is a language and have just memorized procedures that will only solve the problems at hand. As a result, they have developed a very negative feeling towards the subject. This book teaches you the language and concentrates on teaching the understanding. I hope this Chapter has increased your understanding, confidence, and liking for problem-solving and the subject.

I have ended this chapter with some other examples to demonstrate the understanding to be gained by breaking the problem down into manageable parts and only having need to use your understanding of the Two single digit grouping Table, power of Ten notation, and the Number system. Take your time to understand every step in the solution.

365 means $3 \times 100 + 6 \times 10 + 5 \times 1$

278 means $2 \times 100 + 7 \times 10 + 8 \times 1$

You are now working with single digits. You need to know your grouping knowledge for single digit numbers (Times Tables). The addition symbol means you are making groups of like things and the multiplication symbol means "groups of". That means you can simplify the problem by grouping the single digits together one at a time, keeping track of the amounts and different sizes, and adding up all the Like Things at the end. Below I've set up the problem to do that,

Simplify 365 × 278

We'll start by multiplying every element on the top by 8 × 1 then by 7 × 10 then by 2 × 100 and add the results to get a final sum.

$$365 = 3 \times 100 + 6 \times 10 + 5 \times 1$$
$$\times\ 278 = \underline{2} \times \underline{100} + \underline{7} \times \underline{10} + \underline{8} \times \underline{1}$$

$8 \times 1 \times 5 \times 1$	$= 40 \times 1$	$= 40 \times 10^0$	40
$8 \times 1 \times 6 \times 10$	$= 48 \times 10.$	$= 48 \times 10^1$	480
$8 \times 1 \times 3 \times 100$	$= 24 \times 100$	$= 24 \times 10^2$	$2,400$
$7 \times 10 \times 5 \times 1$	$= 35 \times 10$	$= 35 \times 10^1$	350
$7 \times 10 \times 6 \times 10$	$= 42 \times 100$	$= 42 \times 10^2$	$4,200$
$7 \times 10 \times 3 \times 100$	$= 21 \times 1000.$	$= 21 \times 10^3$	$21,000$
$2 \times 100 \times 5 \times 1$	$= 10 \times 100$	$= 1 \times 10^3$	$1,000$
$2 \times 100 \times 6 \times 10$	$= 12 \times 100 \times 10$	$= 12 \times 10^3$	$12,000$
$2 \times 100 \times 3 \times 100 =$	$6 \times 100 \times 100 =$	6×10^4	$\underline{60,000}$
			$101,470$

Every time I have a package of ten, it goes to the next larger size.

$$278 \times 365 = 101,470$$

Using What, How, and Why thinking enables complete understanding of every step taken in the solution process. We have used the language of Mathematics. One symbol, word, or sentence has only one meaning which makes the communication Clear, Correct, and Concise.

CHAPTER 7

Understanding Simple Division

The topic of Division offers an excellent opportunity for using scientific thinking and reasoning, which are some of the attributes of Mathematics.

In this Chapter we are going to explore how What, How, and Why thinking can help you start on the quest towards understanding division. The first step is asking yourself, "What is division, what is it asking you to do?" It's not difficult to come up with a Concept you are familiar with:

- sharing several sweets (The total Number)
- sharing among a number of friends (The number of groups),
- making sure each friend has the same quantity (Number).

The number of sweets each person in the group got, would depend on the number of sweets, and the number of people in the group.

You reflect. You're talking about:

- A quantity of things
- The number of groups you will have.
- The quantity in each group.

From your experiences, and visualization, you can see that you were always sharing out identical quantities. We can now formalize the Division Concept.

Division Concept – Separate a quantity into groups of identical value.

We have formalized the Concept for Division, our focus is now directed to formalizing the Division Rule. We visualize and ask ourselves what do we know about processes that work with "groups of things"? **Eureka!** The topic Multiplication, Chapter 6.

In Chapter 6, we learned the Concept and Rule for Multiplication:

Multiplication Concept- The multiplication symbol "×" means "groups of". You are grouping quantities together

Multiplication Rule – Repetitive addition. You're adding identical value groups together, a given number of times.

Mathematics is a language, so for brevity and clarity, we will use simple symbols, and words, which have only one meaning. We'll use the international names "**Divisor**" for the number of groups we're going to have, **"Dividend"** the quantity, which is being divided, and **"Quotient "**for the amount each group would have.

Remember "×" means groups of. In the Arial font that the book is using, there is very little difference between the multiplication symbol × and the alphabet symbol x so you must be careful you don't confuse the two.

We need to have a symbol for Division that can help us to separate, and keep track of, our three terms **Divisor, Dividend, and Quotient** while we are using division. We will look at a very simple division problem and see if that will give us more clarity.

If we divide 12 by 3, what would we get? We have a line which signifies division the fraction line (We will investigate subtraction in Chapter 8). We will use that symbol.

$$\frac{12}{3} = 4$$ 12 would be the **dividend**, 3 would be the **divisor** and 4 the **quotient**

Dividend

This gives us the relationship $\frac{\textbf{Dividend}}{\textbf{Divison}}$ = **Quotient** but this symbol left no room for the working out.

A better symbol was needed. The Division Symbol: ⌐ was created.

Quotient

Divisor | **Divident**

We may better understand division if we use an actual division problem. We will use the new division symbol and International Symbols. We also need our Double, Single digit, Grouping Table.

Two Single Digit Grouping Table

×	1	2	3	4	5	6	7	8	9	10
1	1	2	3	4	5	6	7	8	9	10
2	2	4	6	8	10	12	14	16	18	20
3	3	6	9	12	15	18	21	24	27	30
4	4	8	12	16	20	24	28	32	36	40
5	5	10	15	20	25	30	35	40	45	50
6	6	12	18	24	30	36	42	48	54	60
7	7	14	21	28	35	42	49	56	63	70
8	8	16	24	32	30	48	56	64	72	80
9	9	18	27	36	45	54	63	72	81	90
10	10	20	30	40	50	60	70	80	90	100

Given 2436 How many groups of 6 does 2436 contain? We will us use our Quotient

Quotient

new Division symbol 6 | 2 4 3 6

We first need to know, how many, and what sort of groups 2436 is made up of? Remember the symbol × means groups of.

2436 is made up of four discrete parts: 2×1000 + 4×100 + 3×10 + 6×1

The largest Size in our number is THOUSANDS so when we divide, we are finding out how many groups of:

- 1000s of 6
- 100s of 6
- 10's of 6
- single digits of 6 Our Number contains

For 2436 to contain any groups of six it must have at least:

- 6000 to contain 1000×6
- 600 to contain 100×6
- 60 to contain 10×6
- 6 to contain 1×6

This pattern is followed in any division problem, so it's very important to understand and remember.

Let's start the division 6 | 2 4 3 6

We are finding out how many groups of 6 the number 2 4 3 6 contains. The largest size in our number is 2000.

Does 2000 contain any 1000×6 ? **No!** we need at least 6000 in order to have any thousands of groups of 6. Because we are at the beginning of the number, we don't place a zero at the front of the number.

The next size is hundreds, does it contain any 100×6 ? **Yes,** it contains 20 from the 2000 and 4 from the 400, that's a total of 24. From our Two Single Digit Grouping Table we know there are 4 groups of 6 in 24. We put 4 in the hundreds place in our Quotient.

$$
\begin{array}{r}
4 \\
6\,\overline{\smash{)}\,2\,4\,3\,6}
\end{array}
$$

We've used up 2400 of our number from 400×6, we need to know how much of the number we have left, so we subtract the 2400 from 2436.

$$
\begin{array}{r}
4 \\
6\,\overline{\smash{)}\,2\,4\,3\,6} \\
-\ 2\,4\,0\,0 \\
\hline
3\,6
\end{array}
$$

We have 36 of our number left

The next size is Tens, does it contain any 10×6 ? **No,** there are only 3 Tens left, we need 60 in order to have 1×10×6. We need to put a zero in the Tens place of the Quotient.

$$
\begin{array}{r}
4\,0 \\
6\,\overline{\smash{)}\,2\,4\,3\,6} \\
-\ 2\,4\,0\,0 \\
\hline
3\,6
\end{array}
$$

Notice we keep the same size placement as in the quotient. We must ALWAYS do this to accurately show the correct value of the quotient.

The next size is the Ones, does it contain any 1×6 ? **Yes!** We have 36, that's 6×6

How did we know that? We used the Table of the values obtained by the groupings of two Single Digit Numbers Table. We must place a 6 in the Ones place of the Quotient and subtract the 36 away from our number to see how much of the number is left.

$$
\begin{array}{r}
406 \\
6\overline{\smash{\big)}\,2436} \\
2400 \\
\hline
36 \\
36 \\
\hline
\end{array}
$$

There is nothing left, were finished. We have found there are exactly 406 groups of 6 in 2436.

$$406×6 = 2436$$

We could easily check our answer by doing the multiplication.

In Chapter 6 we found that the pattern used in the Number System was a reproducible robust pattern. The pattern could be easily modified to give Tables for the grouping of any combination of numbers. The idea that we could use that pattern to separate numbers into a single digit and index 10 number was a very good idea. Multiplication is grouping ideas together and we share those ideas of grouping in Division.

Division is different because we are always working with three ideas:

- Dividend.
- Divisor, and
- Quotient

Let's return to a simple: single digit problem:

$$
\begin{array}{r}
8 \\
7\overline{\smash{\big)}\,56} \\
\end{array}
$$
Eight equal pieces created by dividing 56 by 7.

The Dividend, Divisor, and Quotient were all obtained from a very important and useful Table. The Groupings of Two Single Digit Numbers Table, a Table which we should know and have memorized!

If we wanted to have a **quotient** of 9, which is the largest **quotient**, for a **divisor** of 8 what size **dividend** would we have? Again, we can use our Two single digit Multiplication Table. The Table tells us that largest dividend would be 72. Using our division symbol we would represent our division procedure as follows:

$$
\begin{array}{r}
9 \\
\hline
8 \, | \, 7\,2
\end{array}
\qquad \textbf{Quotient}
$$

Example: 8 | 7 2   **Divisor** | **Divident**

Notice in the **quotient** position the 9 is directly over the 2, the Ones position in the **dividend**. This is important to understand. The **quotient** is telling us the number of groups of the **divisor** in the **dividend**, 9 represent 9×1 (nine groups of one) so the 9 must be in the Ones position in the **quotient**. This correlation must always hold, so that there is no doubt what size the **quotient** represents. What have we learned so far?

- The **quotient** tells you how many groups of the **divisor** are in the **dividend**.
- The number sizes in the **quotient** ALWAYS line up with the same number sizes in the **Dividend**.
- The **dividend** is the total quantity which is being divided.
- For division involving single digit divisors, the Two Single Digit Grouping Table can easily help you derive the value of the **Quotient, Dividend**, and the **Divisor**.
- For division, where the **divisor** is two or more digits, a simple new Table can be easily made which would have the grouping of the **divisor** for 1×**divisor** up to 9× **divisor**.

So far, our ideas work very well with single digit divisors, and we have been able to use our Two Single Digit Multiplication Table. We do have a problem as division problems can involve any size **dividend**, **divisor**, and **quotient**.

We'll need a Table for Tens, Hundreds, Thousands, Ten Thousands etc. Our Two Single Digit Multiplication Table is very robust and the ideas, understanding, and pattern are transferable. Modifying that Table would be a very good start. We'll stick with, and build on, what we know and understand which is our Two Single Digit Grouping Table.

Remember that, in Ariel, there is very little difference in appearance between the multiplication symbol and the alphabet letter x. Starting with the:

Two Single Digit Grouping Table

×	1	2	3	4	5	6	7	8	9	10
1	1	2	3	4	5	6	7	8	9	10
2	2	4	6	8	10	12	14	16	18	20
3	3	6	9	12	15	18	21	24	27	30
4	4	8	12	16	20	24	28	32	36	40
5	5	10	15	20	25	30	35	40	45	50
6	6	12	18	24	30	36	42	48	54	60
7	7	14	21	28	35	42	49	56	63	70
8	8	16	24	32	30	48	56	64	72	80
9	9	18	27	36	45	54	63	72	81	90
10	10	20	30	40	50	60	70	80	90	100

Looking at the Table above we notice that for each value of the divisor we have nine possibilities of the dividend. We also notice that when the dividend become 10, which is 10×1, the quotient for each corresponding value also becomes ten groups larger. For example:

×10	10	20	30	40	50	60	70	80	90	100

The pattern for the ×1 repeats itself. The Two Single Digit Grouping Table has been shown to be a robust, consistent and transferable Pattern. That means:

- If one divisor is increased by a factor of Ten and all its corresponding Quotients increase by Ten. Then the same pattern would occur for all the other divisors.
- If that works for 10s it will work the same for 100s, 1000s, 10000s …
- It also means that this pattern will work for any size divisor.

That's worth investigating. We'll do that below:

$$\frac{9}{6\;|\;5\;4}$$

Example: 6 | 5 4   **Divisor** | **Divident**   Quotient above 9, **Quotient** label

If we increase the value of divisor 6 by a factor of 10, look what happens:

×	1	2	3	4	5	6	7	8	9	10
6	6	12	18	24	30	36	42	48	54	60
10×6	60	120	180	240	300	360	420	480	540	600

Eureka! "increase the divisor by a factor of 10 and the dividends increase by 10 too". That means we can also use the same pattern of the single digits for dividing the single digit into any size dividend. Let's try dividing 2744 by 6

```
      4
6 | 2 7 4 4
  - 2 4 0 0
    -------
      3 4 4
```

First 2744 Represents 2×1000 + 7×100 + 4×10 + 4×1 We start the problem by asking how many thousand groups of 6 are in 2744. To have a thousand × 6 you must have at least 6000 There is only 2000. You then look at the 100 size 2744. There are 27 hundred in 2744. That's enough for 400×6. 4 must be placed in the size 100 place of the quotient and the 2400 subtracted from the dividend. We must subtract the 2400 to see what is left. We are left with 344. The next size of the dividend is the Tens and we have 3×10 in 344.

The Table is versatile, multiplying by a factor of Ten each time allows us to use our existing Table of the single digits. It will allow us to accommodate the different size divisors and dividends. We would only have to multiply the divisor and their accompanying dividends by the same factor of Ten and we would have a new Table for that size divisor. Below is the Table that would be used for each new divisor and each new dividend. Notice all that has been done to our original Table is maintain the same factor of Ten for that grouping exercise. For example:

Divisor	Dividend	Divisor	Dividend
1×6	1×6 = 6	10×6	10×6 = 60
2×6	2×6 = 12	12×6	20×6 = 120
3×6	3×6 = 18	18×6	30×6 = 180
4×6	4×6 = 24	24×6	40×6 = 240

The same pattern continues. It's stable and robust.

ONES, TENS, HUNDREDS, THOUSANDS,
GROUPING TABLE FOR DIVISOR SIZE 6

Ones	Tens	Hundreds	Thousands
1×6	10×6	100×6	1000×6
1×6 = 6	10×6=60	100×6=600	1000×6=6000
2×6 = 12	20×6=120	200×6=1200	2000×6=12000
3×6 = 18	30×6=180	300×6=1800	3000×6=18000
4×6 = 24	40×6=240	400×6=2400	4000×6=24000
5×6 = 30	50×6=300	500×6=3000	5000×6=30000
6×6 = 36	60×6=360	600×6=3600	6000×6=36000
7×6 = 42	70×6=420	700×6=4200	7000×6=42000
8×6 = 48	80×6=480	900×6=4800	8000×6=45000
9×6 = 54	90×6=540	900×6=5400	9000×6=54000

You can see the influence of our original single digit grouping Table. The pattern is reproducible, robust, and continues for the next size which is 10000 and larger sizes. Let's construct a Table for any value divisor. Let N = Any real number

Ones	Tens	Hundreds	Thousands
1×N	10×N	100×N	1000×N
1×N = 6N	10×N=60N	100×N= 600N	1000×N=6000N
2×N = 2N	20×N=120N	200×N=1200N	2000×N=12000N
3×N = 3N	30×N=180N	300×N=1800N	3000×N=18000N
4×N = 4N	40×N=240N	400×N=2400N	4000×N=24000N
5×N = 5N	50×N=300N	500×N=3000N	5000×N=30000N
6×N = 6N	60×N=360N	600×N=3600N	6000×N=36000N
7×N = 7N	70×N=420N	700×N=4200N	7000×N=42000N
8×N = 8N	80×N=480N	800×N=4800N	8000×N=45000N
9×N = 9N	90×N=540N	900×N=5400N	9000×N=54000N

Being able to substitute a symbol in a new pattern to represent any number and have the pattern work with the symbol is a very powerful mental tool for testing the robustness, reproducibility and reliability of the pattern. It demonstrates the power of Algebra. Let's try out our new ideas with a more difficult problem. This will demonstrate how even an apparent difficult problem can be easily solved when you understand. You should never be afraid of a number just because it looks big. Work it the same way. We'll start with a Dividend of 849,936 and a Divisor of 234. The first thing we will need to do is build a Grouping Table for our Divisor which is 234. No problem.

We need a reference-Table when we have divisors of two or more digits as we need to know how many groups of the divisor are contained in the corresponding quotient. Our single-digit×234 column is our reference Table. It gives us the 1-9 values of the quotient

Grouping Table for 234.

×234	Single digit 234s	×10×234	×100×234
1×234	= 234	= 2340	= 23400
2×234	= 468	= 4680	= 46800
3×234	= 702	= 7020	= 70200
4×234	= 936	= 9360	= 93600
5×234	= 1170	= 11700	= 117000
6×234	= 1404	= 14040	= 140400
7×234	= 1638	= 16380	= 163800
8×234	= 1872	= 18720	= 187200
9×234	= 2106	= 21060	= 210600

Given: 234 $\overline{)8\,4\,9,9\,3\,6}$

How many groups of 234 are there in 849653?

We are talking about groups of 234, we are talking about a three-digit number.

We would need a three-digit number to contain a three-digit number.

Looking at our dividend, the first three digits of our number would give us 849,000.

We therefore are asking how many 1000×234 are in our number?

We look at the quotients in the single digit×234s column.

The divisor tells us that the largest number that is less than 849 and contains ×234 is 702.

We are talking about ×1000s, so there are 3000×234 in 849653.

Notice- This is exactly the same procedure we followed for a much smaller and simpler division problem. We are now ready to start the division process.

```
            3 0 0
234 | 8 4 9 9 3 6
    - 7 0 2 0 0 0
          9 3 6
```

```
          3 0 0 4
234 | 8 4 9 9 3 6
    - 7 0 2 0 0 0
          9 3 6
        - 9 3 6
```

To clarify: There are 3000 groups of 234 in our dividend, so we place a 3 in the thousands place in our quotient, directly above the thousands place in the dividend, and subtract the 702000 from our dividend. We have 936 left which is made up of 9×100 + 3×10 + 6.

We look at the 936 does that contain any 100s of × 234? **No** it's too small, same for the 93×10. we place a 0 in the 100s place and a 0 in the 10s place. For the Ones we have 936×1 and our Quotient Table tells us that's enough for 4×234. We place a 4 in the Ones place of the quotient and subtract the 936 from what is left of our number. We have nothing left. We are finished with the division process. **Congratulations**. You have determined, using simple ideas, Concepts, Rules, logic, visualizations, patience, and determination to find the number 849936 contains 3004 groups of 234. You have been able to persevere, think and correctly put a lot of ideas together for a significant length of time. That capability will serve you well in everyday life situations.

Following is another example problem for reinforcing what has already been learned. It is a good self-confidence building step.

```
35 | 245678
```

You are working with a 2-digit divisor. Your need a reference Table. No problem. We'll put it to the left where we need it. You start by asking how many groups of 35 are in 245,678. 35 is a two-digit number so we are only interested in the first two digits

Dividend	Quotient
1×35	35
2×35	70
3×35	105
4×35	140
5×35	175
6×35	210
7×35	245
8×35	280
9×35	315
10×35	350

of the dividend that gives us 240000 or 24×10,000. It has no ×35×10,000. We then look at 240,000. **Yes,** 240,000 contains 6×35×1000 which is 210,000. We place 5 in the 1000s place of the quotient and subtract 210,000.

We have 5678 of our number left. We look at the first two digits. That's 56×100. From our Table we see that 56×100 contains 1×3500. We place a 1 in the 100s place and subtract 3500 from the remainder. We look at the remainder and see how many Tens it contains. We look at the first three digits of the remainder. It contains 217×10. Our Table tells us it contains 6×35×10. We place a 6 in the Tens and subtract 2170. We are left with a remainder 8. We need 35×1 for the remainder to have 1×35×10. We put a 0 in the Ones place in the quotient and "Remainder" in front of the 8 to let everybody know that there is a remainder of 8 with the answer.

```
              5 1 6 0
   35 | 2 4 5 6 7 8
      - 2 1 0 0 0 0
          5 6 7 8
        - 3 5 0 0
          2 1 7 8
        - 2 1 7 0
                8
```

Important! When you start a division process you are always asking: How many groups of the divisor go into the dividend. YOU DON'T START BY SAYING "IT DOESN'T GO". This only confuses the brain because when you say this, you are only talking about the first digit of the dividend not the **whole** dividend. It should ALWAYS be:

How many groups of the divisor go into the whole dividend.

There was a lot of effort spent in devising this method for solving division problems; but a lot was learned through gaining practice in logical thinking, reasoning and scientific thinking.

There is no understanding gained by in a rote manner. memorizing the standard method taught for division. In fact, you will not be able to get a job anywhere, because you are proficient in doing the rote learned procedure. They will politely tell you that we use calculators here or tell you the cash register will do those sums faster and more accurately. The real value in investigating Division is : the patience, and tenacity gained; from thinking and reasoning; manipulating Concepts, and Rules, logically and scientifically, while problem solving. This is how you develop your ability to think at a higher level. These are the skills you need, to successfully solve the problems you are encountering now, and in the future.

For consolidating what you have learned, I will perform the actual steps required to do the problem but will leave writing the method which explains the reasoning behind each step in the solution, for you to do.

Dividend	Quotient
1×84	84
2×84	168
3×84	252
4×84	336
5×84	420
6×84	504
7×84	588
8×84	672
9×84	756
10×84	840

Divide 363,741 by 84

```
            4 3 3 0
      84 | 363,741
         - 336 000
           27 741
         - 25 200
            2 541
          - 2 520
Remainder     2 1
```

Many times, when you have a remainder you want the remainder as a decimal fraction. No problem. The decimal part is still part of the groups of 10. You don't want to change the value of the dividend you only want to change the remainder sizes into negative index sizes. You only have to add zeros for the decimal places you want and continue the division process. This is demonstrated below

Divide 363,741 by 84

$$
\begin{array}{r}
4330.25 \\
84\overline{\smash{)}363741.00} \\
-\ 336000.00 \\
\hline
27741 \\
-\ 25200 \\
\hline
2541 \\
-\ 2520 \\
\hline
21.0 \\
16.8 \\
\hline
4.20 \\
4.20 \\
\hline
\end{array}
$$

For consolidation I shall demonstrate a division involving decimals. Remember we are following patterns and the patterns we have been using are universal.

Divide the following.

$$
3.2\ \overline{\smash{)}0.2664}
$$

We notice the divisor has a decimal point. This only complicates the Grouping Knowledge Table. We can easily eliminate the decimal by using the pattern of groups of ten. Each time you multiply by ten the value of the quantity increases by a factor of ten. We have already been using that pattern. We want the decimal to move to the right one place in the divisor, so we multiply by 10. To maintain the pattern, we must multiply the dividend by the same factor. This means we have maintained the same ratio. We will do much more of this when you start the Fraction and Equation sections in Chapter 8. Moving the decimal the same number of places in the divisor and dividend we have:

$$\frac{0.2664}{3.2} = \frac{0.2664}{3.2} \times \frac{10}{10} = \frac{2.664}{32}$$

> We have moved the decimal point one place in the divisor increasing its value to 32 without changing the value of the problem

We are working with a minus index of 10 going from 10^{-1} to 10^{-2} to 10^{-3} to 10^{-4}

Multiplier	10^{-1}	10^{-2}	10^{-3}	10^{-4}	
× 32	.1	.01	.001	.0001	0.08325
1 × 32 = 32	3.2	.32	.032	.0032	32) 2.66400
2 × 32 = 64	6.4	.64	.064	.0064	2.56000
3 × 32 = 96	9.6	.96	.096	.0096	.10400
4 × 32 = 128	12.8	1.28	.128	.0128	.09600
5 × 32 = 160	16.0	1.60	.160	.0160	.00800
6 × 32 = 192	19.2	1.92	.192	.0192	.00640
7 × 32 = 224	22.4	2.24	.224	.0224	.00160
8 × 32 = 256	25.6	2.56	.256	.0256	.00160
9 × 32 = 288	28.8	2.88	.288	.0288	0

Checking our answer for 2.6 6 4 0 0 0 .08325 × 32 = 2.664 Our answer is correct!

Rules and Concepts needed to Simplify when working with whole numbers or decimals include:

- manipulating the decimal point to make the multiplier a whole number;
- using the single digit grouping Table;
- The Indice Rule - Add the indices when multiplying same values which have an index;
- the Addition Concept and Rule;

- the Subtraction Concepts (there are two Concepts) and Rule; and the
- Number System.

For example: Which Concepts and Rules will you use when you Simplify the following expressions below. Perhaps in the past you have only memorized procedures to Simplify some of the expressions. It's important you recognize and understand the Rules and Concepts being used during problem-solving. They are the mental tools that you use in any logical problem-solving situation. This habit you will develop enables you to increase your ability to think at a higher thought level.

$8 \times 32 = 8 \times 2 + 8 \times 3 \times 10 = 16 + 24 \times 10 = 16 + 240 = 256$

$.0003 \times 0.4 = 3 \times 10^{-4} \times 4 \times 10^{-1} = 12 \times 10^{-5} = .00012$

You will have much more opportunity to put ideas together using Concepts, Rules, and Patterns (scientific thinking) in Chapter 8 and 9

You are practicing, making sure there is an appropriate rule to follow, and you are following it. You've used the Power of Ten Rule; The Index Rule, Adding the indices when multiplying same values which have an index, and single digit grouping Table.

With the level of understanding gained by using ideas in an organized logical manner you are ready for Chapter 9 where you will get practice in problem-solving that is not so number focused, but more concerned with abstract ideas.

You now have the opportunity of setting up you own problems and being able to check your answer. It's important that you take care and responsibility for your own learning needs. You need to be saying to your -self, as you do a problem. "What am I doing", "How am I going to do it", and checking "Why am I doing it" This habit ensures that you are always in charge of your thinking and developing self-confidence in your ability as a problem-solver.

CHAPTER 8

Understanding the Algebraic Variable, Fractions, and The Equation,

Mathematics is a unique language. It communicates Clearly, Correctly, and Concisely. It does this using algebraic symbols, variables, Concepts and Rules. The symbol is usually the first letter of the word that it is standing for. It makes the beginning sound of what you are representing, easy to use and remember. This makes sense. It's important to understand a lot of words could start with the same alphabet letter, so it is very important the symbol is clearly defined.

Numerals are algebraic symbols. For example, numeral 3 stands for a quantity of something, it's universally understood. It doesn't make any difference what you are referring to . If you only have three of them, we know exactly how many you have of them - there is no confusion. If you were working with Force, you would use capital F. In physics, F always stands for force, it's universally accepted, therefore it's an algebraic symbol. Lower case m for mass, and a, for acceleration, they are associated with F. They are also used universally; they would be called algebraic symbols too.

When the same symbol is used in several entirely different situations with a meaning only associated with that situation, the symbol is then called an algebraic variable. In this case the symbol must be defined and re-defined when the meaning changes. The symbols x and y would be good examples. They are frequently used in equations where the value is unknown but is to be determined. When the symbol is called an algebraic variable, it's a reminder for you to check to make sure you are using the correct meaning, and that it's clearly defined.

There are four main operations when working with numbers. Addition, Subtraction, Multiplication, and Division. Because Mathematics is logical and the language communicates clearly, correctly, and concisely, the rules of operation are very precise and must be carefully followed. We always try to minimize the number of symbols needed for any operation. The +, −, and ÷ are always used; not so for the multiplication (grouping symbol). The Multiplication × is not required when working with algebraic variable and numbers. There is no × symbol between variables or between numbers and variables; the number is always in front; alphabet variables are in alphabetical order. For Example:

- 3 × 4 means you have three groups of four, 4 + 4 + 4
- 3abc means you have 3 groups of abc ; abc + abc + abc
- Both 3 × 4 and 3abc have discrete values. That means if you had a pile of them, you could arrange them in any order, and they would always be the same quantity, and the same value would be clear to everyone.

When we communicate in Mathematics, especially verbally we must always be clear, correct, and concise. If I said the amount is 3adf it makes a specific sound, but if a second person said the amount was d3fa and a third person said the amount was fda3 you would be confused as to what the correct amount was. All the values sound different. You would no longer say that the language is clear, correct, and concise. This is why the number is always in front; and alphabet variables are in alphabetical order. When there is no instruction symbol mentioned the × symbol is there, but invisible. For example: Given a = 2, b = 3, and c = 4

- a + b + c = 9, b + c + a = 9, c + a + b = 9 a + c + b = 9
- abc means a × b × c = 24; b × a × c = 24; c × b × a = 24. You can arrange the quantity in any order using a multiplication or addition process, and you'll still have the same amount. Visualize – you have not added or taken any away, just changed the positions of the individual ones. If you had a bag of money, will the amount inside change if you gave the bag a shake making the coins move around?

- Using abc again $\frac{a}{c} = \frac{2}{4}$ = 0.5 But rearranging we have $\frac{c}{a} = \frac{4}{2}$ = 2

  You will change the result by inverting the division process. These are two different processes

  Using abc again c – a = 4 – 2 = 2 reversing the subtraction you'll get a – c = 2 – 4 = –2 The actual amount stays the same, but you change from what you had to what you owe!

Symbols, Fractions, and Equations all work together as a functioning, well organized team. You'll learn more about this later when we enter the section on Equations.

As mentioned previously Mathematics is a problem-solving, logical, scientific language. During problem-solving, a symbol, word, or sentence has only one meaning. In Chapter 6 we found when following a multiplication instruction, we could read the instruction from left to right or right to left without changing the value of the result.

English is quite different, for example you are taught from the very beginning that you always read from left to right when reading or writing. Surprisingly this is not always true. If it's more convenient or makes more sense to read from right to left, your brain will take over, and without realizing it. you'll read it right to left. As a very simple example could you softly, read out the following: "You will need $5 for the bus fare". You will be reading left to right for sure, but I bet you didn't read $5 from left to right. As well, you probably didn't realize that $ and 5 are universal algebraic symbols. Instead of writing dollars each time you used a simple, easy to recognize symbol, as well, the symbol 5 represents five of something.

Another example of how the symbols can be read is the following. Suppose in our previous fruit shop example in Chapter 6 you wanted to say something about the algebraic variable you used. You would put a small symbol at the foot of the variable and in front of it. It's called a **subscript**. For example: You would first define the symbols. Let A = Apple, B = Banana, R = Red, Y = yellow, and G = green.

You have put the following fruit in your bag. $3A_R + 7B_G + 8A_Y + 6B_Y$ (pronounced 3A sub R, 7B sub G, 8A sub Y and 6B sub Y). What you have written is an algebraic expression; you know that in Mathematics you communicate Clearly, Correctly, and Concisely. You also know that the brain is a problem-solving, logical, scientific organ. Speaking in English, how would you tell someone, what you have put in your bag ?

There's a good chance you started off, after a very short pause, saying "Three red apples". After you heard Three red apples, you would then quickly continue with, seven green bananas, eight yellow apples, and six yellow bananas. You would have recognized the pattern from just hearing five red apples. Some of the specialty skills our brain has is the ability to problem-solve, recognize patterns, compare previous patterns with the present and modify them to fit the present problem-solving situation.

Your brain is active, and problem-solving 24/7, it doesn't sleep. While you're asleep, it's analysing the information from the sounds you hear, movements around you, your breathing etc. If it feels you're in danger, you'll probably wake up with a start.

The problem-solving we have at hand is communicating the different fruit you sold. You've decided to use algebraic variables. You'd first have to define them, Let F = fruit etc. We haven't changed the definitions of our previous variables. You now decide to purchase flowers. You decide to use the sound of the symbol F again. You would use lowercase f = flowers. In this case F would still stand for fruit. If you change the meaning of the symbol F then you'd have to redefine them.

Let f = flowers. You purchase $10f_Y$ and $8f_R$. How would you communicate this in English? If you were thinking ten yellow flowers and eight red flowers, that's logical.

If you are working in a rote memory daze, you'd possibly think Ten flowers that are yellow and eight flowers that are red. However, if you were thinking naturally your brain would be thinking logically.

Tomorrow you are going to buy apricots. Using algebraic variables allows the brain to be more efficient, not having to work with a lot of unnecessary, distracting, information which is called information over-load. You would then define a = apricot. You have not redefined A, so capital A still means apples.

Once you realize you're only using common sense working with algebra, it's convenient and simple. You're only using simple symbols that stand for a single word, sentence, or paragraph, and it makes it easier to think and communicate ideas. You'll use algebra whenever you feel like it. The use of Subscripts and Superscripts have already been briefly covered in Chapter 2 but will be discussed more thoroughly in this chapter.

Fractions

When working with Fractions we will be using all the following mental tools:

You were introduced to the **Three One Rules (TOR)** in Chapter 6. You'll see how powerful, and useful they are for manipulating fractions and equations.

Concepts and Rules, and Concepts and Rules are facts and mental tools.

The **Concept Always** tells you what to do, and

The **Rule Always** tells you how to do it.

The Addition Concept – Make groups of 'Like Things'

The Addition Rule – Only put Like Things together.

Subtraction Concept – The Subtraction symbol means How much you owe, or how much you take away, and you only take Like from Like.

Subtraction Rule – You are taking away from what you have or from what you owe.

You first check to see if there is enough for you to take away from. If not, you are in a "taking away from what you owe situation". What you owe must be on the Top. Review Chapter 5.

Equivalents Rule – Equivalents can always be substituted for Equivalents. For example: 2×3 substituted for 6; 3×4 substituted for 4×3; two fifty cent pieces for $1 etc. Equivalent is defined as equivalent in value or meaning. **The word "equals" is not a mathematical expression.** It has many meanings and therefore doesn't fit the bill for clear, accurate, and concise. It creates confusion and misunderstanding.

The Three One Rules (TOR): They can be used left to right or right to left. They're very important; Using TOR1 you can eliminate multiplication from a problem if it makes the problem easier to do OR insert multiplication if it will make the problem easier. In both cases you can only do the above providing the overall effect is only multiplying the whole problem by one. For TOR 2 you can do the same with division, eliminating division, or adding division, PROVIDING the overall effect is only dividing the whole problem by one. Looking at TOR's ideas from left to right and right to left we have:

1. Anything multiplied by One remains the Same in value. Anything remains the same in value if it is multiplied by One.

2. Anything divided by One remains the same in value. Anything stays the same in value if it is divided by One.

3. Anything divided by itself is One.

One is equivalent to something divided by itself (Exclude divide by zero. If you divide by zero it's equivalent to asking. What's the largest number? There's no such number. The infinity symbol "∞" is used to represent this situation).

Working with equations requires you to understand fractions so you are confident in their correct and proper use during problem-solving.

Fractions

A Fraction can perform two compatible, interchangeable, convenient, functions. The first meaning is being able to divide a whole entity (a pie, cake, a box of apples etc.) into equal fractional parts. These parts are referred to as fractions. In this situation:

The Fraction Concept – Divide the whole entity 1, into equal fractional parts.

The Fraction Rule – Divide the Numerator (top) by the Denominator (bottom).

A Fraction has the following format: $\frac{x}{y}$ where the line is an instruction line to divide the top, the numerator, which has a value of 1 signifying the whole thing, and by the denominator y which has any value but tells you how many fractional parts. For example: half a box (1/2), two thirds of a box (2/3), three quarters of a box (3/4) etc.

A fraction can also stand for "percent" for example: Instead of half a box you could say 50 percent of the box. We are still using the idea $\frac{x}{y}$ but instead of having a numerator equal to 1, the denominator becomes equal to 100 without changing the value of $\frac{x}{y}$. All that is required right now is to know Percent means the number of hundredths.

$\frac{x}{y} = \frac{1}{2} = 50$ Percent $= \frac{50}{100} = 50$ Percent of the article. Percent is used MUCH MORE frequently than saying the fraction of the article. Percent will be thoroughly discussed a bit later.

The Fraction Format $\frac{x}{y} =$ has another very important use. It is a very simple and convenient Format when used as a division instruction, and even more powerful when division instructions are multiplied together or divided by each other.

Let's take another look at the division symbol and the Fraction symbol:

$$\text{Divisor} \overline{\left)\underset{\text{Divident}}{\overset{\text{Quotient}}{}}\right.} \quad \text{and} \quad \frac{x}{y}$$

$$\frac{\text{Dividend}}{\text{Divisor}} = \text{Quotient} = \frac{x}{y}$$

This means that we can use the Fraction symbol for $\frac{\text{Dividend}}{\text{Divisor}}$ and the result will be the quotient

Fraction Rule - The line between them means divide the Numerator (top) by the Denominator (bottom).

Multiplication Rule for Fractions - Multiply **Numerators** together, and the **Denominators** together AFTER first checking to see if you can do any division.

You've built up a good kit bag of Concepts and Rules, so you're going to be able to do some high-level thinking by the time we're done.

Let's start putting all the above ideas into action. Remember we are looking for patterns, something that keeps occurring consistently. If the procedure or observation continues to repeat itself, we investigate to determine if it is based on facts. If so, it becomes a Rule and a dependable mental tool.

We have shown that a Fraction has interchangeable Functions suitable for doing division.

Let's first look at a fraction using concrete symbols to reinforce some of our abstract ideas.

Fraction Concept - You are dividing the numerator which has value a of "1" (The whole) into equal fractional sizes.

In this situation the format of the fraction is the follows:

Fraction Rule – You divide the Numerator, by the Denominator which designates the number of equal parts required.

Let's start with something you're familiar with, a pie. We'll represent the pie by a circle.

We want to divide the pie into two identical pieces. We can represent this with a fraction because a fraction is an instruction to divide.

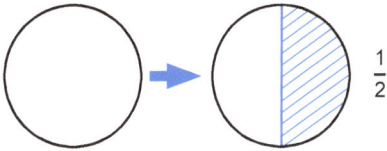

If we reflect and visualize. The Numerator represents the whole pie and the Denominator represents the Number of equal pieces we obtain. We have one piece of size $\frac{1}{2}$

$\frac{1}{2}$

What we have just described is a universal idea as you can clearly visualize the two equal pieces. If we divided the whole into four equal pieces, we still have the clarity of what the fraction represents.

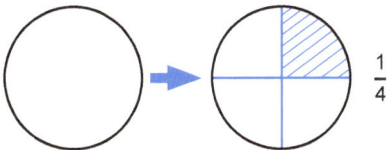

We can now generalize this pattern. We have divided the Numerator into 4 equal parts We have one piece of size $\frac{1}{4}$

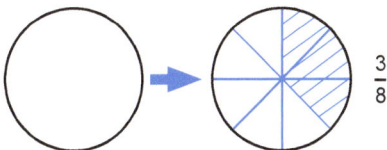

$\frac{1}{4}$

This means you have 3 equal pieces of size $\frac{1}{8}$

$\frac{3}{8}$

If we want to divide a 20-gram pie (Dividend) into 4 equal pieces (Divisor) we can use a fraction symbol as an Example:

Divide the Numerator by the Denominator $\dfrac{20 \text{ grams}}{4} = 5 \text{ grams}$

We get four equal pieces, each of value 5 grams $5 + 5 + 5 + 5 = 20$

We now have a new Fraction idea: The Mathematical terms for fractions

- The NUMERATOR tells you the TOTAL QUANTITY (**Dividend**) being divided
- The DENOMINATOR (**Divisor**) the NUMBER OF EQUAL PIECES being created and
- The QUOTIENT is the RESULT OF THE DIVISION giving us the value of each piece.

Comparing division and Fraction terms. We had the Divisor, the Dividend, and the quotient.

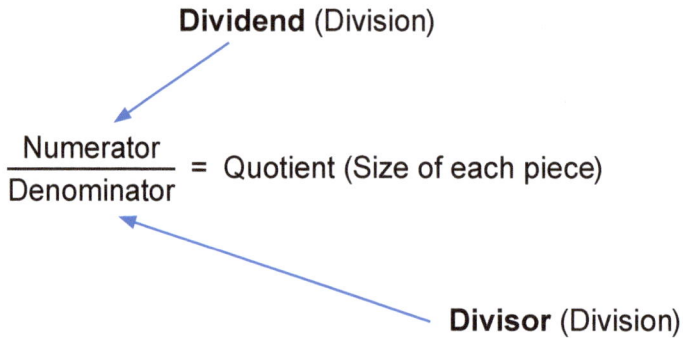

Dividend (Division)

$\dfrac{\text{Numerator}}{\text{Denominator}} = $ Quotient (Size of each piece)

Divisor (Division)

Let's investigate addition of Fractions (Multiplication expressions)

Addition Concept – Make groups of Like Things.

Addition Rule – Only put Like things together

Let's look at a typical Addition of Fractions problem.

Given: $\dfrac{5}{16} + \dfrac{5}{8}$

It's an ordinary looking addition of fractions problem. What are the fractions telling us about themselves? The first fraction says, divide the Numerator into 16 equal pieces. The second fraction is saying divide its Numerator into 8 equal pieces. The addition symbol says make groups of Like Things. To be able to follow the Addition Rule we must make all the sizes the same. The fractions become the same size when all the Denominators are the same size. We must do this WITHOUT CHANGING THE VALUE of the fractions.

Let's reflect for a moment to check that we are thinking this out correctly. We can see that that we have 5 pieces of size 16 and that it would be easy to change the 5 pieces of size 8 to 10 pieces of size 16 using TOR the Three One Rules. If the sizes are the same size. then we can add them together. We can add all of the pieces together now because they are all the same size. Together there would be 15 pieces of size 16. We can see that we have correctly followed the Addition Rule: ONLY put Like Things together.

$$\frac{5}{16} + \frac{5}{8} = \frac{5}{16} + \frac{2}{2} \times \frac{5}{8} = \frac{5}{16} + \frac{10}{16} = \frac{15}{16}$$

Our thinking is correct. We must have a Like Size, and the Like Size must share all the factors that are shared by all the denominators. If we just multiplied the two denominators together, 16 and 8 would give us 16 × 8 = 128. That would give us a number which would contain both denominators, but it might give us a much bigger number than we need. If the denominators had common factors, that would make the Like Size smaller.

Let's see, the denominators are 2 × 8 and 8. We can get rid of one 8, that would give us 2 × 8. That's smaller and still contains, both 8 and 2 × 8. We now have our Like Size, which is divisible by both 8 and a 16. What we have done in this problem is what we would have to do in any addition of fractions problem. Now let's lay the problem out using our algebraic symbols. It should be much easier to follow and understand. For reinforcement of our thinking and reasoning, I'll still include some English explanations. As you become more confident (That means your understanding is increasing) it won't be necessary to include the English explanations.

$$\text{Given: } \frac{5}{12} + \frac{5}{8} = \frac{\text{Quotient}}{\textit{Lowest} \text{ Common Denominator}} = \frac{\text{Quotient}}{\text{LCD}}$$

We have to make the Denominators the Like Size. The Like Size has the same factors as the LCD. This is a bit more complicated.

We now break each denominator down into its component factors. We then make up the LCD, it contains all the common factors the Denominators share. Both fractions have a factor of 4 in their Denominators It makes sense to use LCD for Lowest Common Denominator

$$\frac{5}{12} + \frac{5}{8} = \frac{\text{Quotient}}{3 \times 4 \times 2} \quad \text{We've constructed the LCD. } 3 \times 4 \times 2$$

3×4 2×4   We now proceed to make the Like Sizes

$$\frac{2}{2} \times \frac{5}{3 \times 4} + \frac{3}{3} \times \frac{\overset{5 \times 3}{5}}{2 \times 4} = \frac{10+15}{3 \times 4 \times 2} = \frac{15}{3 \times 4 \times 2} \quad \text{Check, we see division by 3}$$

$$= \frac{5}{4 \times 2} = \frac{5}{8} \quad \text{Notice how many times you used TOR in the solution process}$$

A more complex addition of fractions problem, one that involves having to do multiplication first.

Given: $\dfrac{5}{16} + \dfrac{2}{3} \times \dfrac{5}{8}$ We must simplify $\dfrac{2}{3} \times \dfrac{5}{8}$ First. Notice this problem has a + part and a multiplication part. The multiplication must be done first to comply with the + part.

$$\underset{2\times 8}{\dfrac{5}{16}} + \dfrac{2}{3} \times \underset{2\times 4}{\dfrac{5}{8}} = \dfrac{5}{16} + \dfrac{5}{12} = \dfrac{5}{4\times 4} + \dfrac{5}{3\times 4} = \dfrac{\text{Quotient}}{\text{LCD}} =$$

$$\dfrac{\text{Quotient}}{4\times 4\times 3} = \dfrac{3}{3} \times \dfrac{5}{4\times 4} + \dfrac{4}{4} + \dfrac{5}{3\times 4} = \dfrac{15+20}{4\times 4\times 3} = \dfrac{35}{48}$$

In many school classrooms this problem is solved using rote methods of learning. It enables the learner to simplify the problem with simple procedures. With sufficient practice they becme competent at solving this type of problem. Unfortunately, rote learning is not going to significantly increase their ability to think, reason, and analyse information. It only enables the student to be more efficient in reaching the required answer.

Their lack of understanding means their self-confidence towards competently applying what they have learned towards resolving other types of problems is very low. They are mainly being trained to achieve high marks on the final Assessment, which only contains similar problems to the ones they have been practicing on.

The measure of a student's ability to think and learn at a higher level is measured by the number of ideas that they can manipulate, and work with, in an organized logical manner. (Vygotsky)

This means that a student's thinking and reasoning power is not fixed, such as in IQ, but can be increased by having them reason their problems out using Concepts, Rules, Patterns and Reflection. Introducing the student to more complex problems which require them to work with more ideas will enable them to reach a much higher level of thinking and reasoning

Let's try our new level of understanding, simplifying a more difficult Addition of Fractions problem: I will leave out some of the explanation to give the student more opportunity to think for themselves.

$$\text{Given: } \frac{5}{24} + \frac{7}{18} + \frac{5}{36} + \frac{1}{42}$$

We recognize that we are working with fractions. We know the following: a fraction is an instruction to divide; the Denominator is the size of the fraction; and the Numerator tells you how many groups of the denominator you have.

We can see that the numerators and the denominators are all different amounts and sizes. We know we can only add the numerators together if their denominators are all the same size. Fortunately, we have mental tools, The Three One Rules, which enable us to change the fraction's denominators to become equal to the Like Size, size without changing their values. That Denominator is the same as the COMMON DENOMINATOR.

I call this same size denominator the LIKE SIZE. The LIKE SIZE is the product of the factors the denominators share. It is identical to the COMMON DENOMINATOR. We could construct the COMMON DENOMINATOR by just multiplying all of the denominators together, but the size obtained would be much larger than what we need, and more cumbersome to work with. In the previous problem we found once we express the COMMON DENOMINATOR as product of the denominator's common factors, it was easy to express each of the original denominators as a LIKE SIZE by making sure each one had the same factors as the COMMON DENOMINATOR. This would enable us to construct a much smaller LIKE SIZE only made up of the factors that the Denominators share. The Denominators which I call the LIKE SIZE, commonly called the smallest Common Denominator.

We are now ready to start constructing the LIKE SIZE. We notice all the Denominators are even numbers. All even numbers are divisible by 2, This gives us a major factor of 2. If we divide the denominators by 2 we are left with: 12, 9, 18, and 21. We can see all of these numbers are divisible by 3. This means we have a larger factor of 2*3 which is 6. We can now use our factor of 6. We divide each denominator by 6 to find any other factor. We continue this process until we have all the factors that the denominators are made of. This is shown below:

Putting things together we have:

$$\frac{5}{24} + \frac{7}{18} + \frac{5}{36} + \frac{1}{42} = \frac{Quotient}{Factors\ in\ Denominator} = \frac{Quotient}{6\times2\times2\times3\times7}$$

$$
\begin{array}{llll}
4\times6 & 3\times6 & 6\times6 & 7\times6 \\
2\times2\times6 & & 3\times2\times6 &
\end{array}
$$

LCD

We reflect on what we have done. We found a major factor 6, and minor factors 2×2, 3, and 7. Putting only the common factors together we have 6×2×2×3×7

$$\frac{5}{24} + \frac{7}{18} + \frac{5}{36} + \frac{1}{42} = \frac{Quotient}{6\times2\times2\times3\times7}$$

$$
\begin{array}{llll}
4\times6 & 3\times6 & 6\times6 & 7\times6 \\
2\times2\times6 & & 3\times2\times6 &
\end{array}
$$

The *LIKE* SIZE must contain all of the *COMMON FACTORS* so we check. We start with:

24 = 2×2×6 *we need* ×3×7
18 = 3×6 *we need* ×2×2×7
36 = 3×2×6 *we need* ×2×7
42 = 7×6 *we need* ×2×2×3

We have what we need, to modify the denominators to become the Like Size.

$$\frac{3\times7}{3\times7} \times \frac{5}{24} + \frac{2\times2\times7}{2\times2\times7} \times \frac{7}{18} + \frac{2\times7}{2\times7} \times \frac{5}{36} + \frac{2\times2\times3}{2\times2\times3} \times \frac{1}{42} = \frac{Quotient}{2\times2\times6\times3\times7}$$

Putting things together we have:

$$\frac{5}{24} + \frac{7}{18} + \frac{5}{36} + \frac{1}{42} = \frac{Quotient}{Factors\ in\ LCD} = \frac{Quotient}{6\times2\times2\times3\times7}$$

| 4×6 | 3×6 | 6×6 | 7×6 | | |
| 2×2×6 | | 3×2×6 | | | *LCD* |

We reflect on what we have done. We found a major factor 6, and minor factors 2×2, 3, and 7. Putting only the common factors together we have 6×2×2×3×7

$$\frac{5}{24} + \frac{7}{18} + \frac{5}{36} + \frac{1}{42} = \frac{Quotient}{6\times2\times2\times3\times7}$$

4×6 3×6 6×6 7×6
2×2×6   3×2×6

The *LIKE* SIZE must contain all of the *COMMON FACTORS* so we check. We start with:

24 = 2×2×6 *we need* ×3×7
18 = 3×6 *we need* ×2×2×7
36 = 3×2×6 *we need* ×2×7
42 = 7×6 *we need* ×2×2×3

We have what we need, to modify the denominators to become the Like Size.

$$\frac{3\times7}{3\times7} \times \frac{5}{24} + \frac{2\times2\times7}{2\times2\times7} \times \frac{7}{18} + \frac{2\times7}{2\times7} \times \frac{5}{36} + \frac{2\times2\times3}{2\times2\times3} \times \frac{1}{42} = \frac{Quotient}{2\times2\times6\times3\times7}$$

All of the denominators are LIKE SIZES. We next simplify the Numerators

$$\frac{3\times7}{3\times7} \times \frac{5}{24} + \frac{2\times2\times7}{2\times2\times7} \times \frac{7}{18} + \frac{2\times7}{2\times7} \times \frac{5}{36} + \frac{2\times2\times3}{2\times2\times3} \times \frac{1}{42} = \frac{\text{Quotient}}{2\times2\times6\times3\times7}$$

Simplifying we obtain

$$\frac{21}{21} \times \frac{5}{24} + \frac{28}{28} \times \frac{7}{18} + \frac{14}{14} \times \frac{5}{36} + \frac{12}{12} \times \frac{1}{42} = \frac{\text{Quotient}}{2\times2\times6\times3\times7}$$

$$\frac{105}{504} + \frac{196}{504} + \frac{70}{504} + \frac{12}{504} = \frac{105+196+70+12}{504} = \frac{393}{504}$$

If we had multiplied the individual denominators together, that would always create a LIKE SIZE as it would contain all the individual denominators as factors. Unfortunately, it would be a much larger Like Size than we would like to work with. For example: in this case it would have been 653,184

The problem-solving procedure we have just used is a very powerful useful tool, as well as it being an important pattern to file and remember. We will simplify a similar Addition of Fraction problem but will skip some of the steps you might be able to do in your head. Let's consolidate our knowledge and understanding with the following Addition of Fractions problem.

We can now define the Like Size: The Like Size must contain the largest common factors Major and Minor as well as the other factors so when grouped together contains all the individual denominators. For example:

The Major factor – It's contained in the denominator of each fraction.
The Minor factor– It's contained in two or more of the denominators.
The individual factors – Those numbers left after the major and minor factors.

The problem below is an equation. Equations are in Chapter 9. The only Concepts and Rules you haven't had yet are the:

Equation Concept: The equals sign "=". Means the left side of the equals sign has the identical value as the right

Equals Rule: You can do whatever you like to one side of the equals sign PROVIDING you do the equivalent thing to the other side.

For example: Given that 6 = 6

I decide to multiply the left side of the equation by 6. Today I decided to multiply the left side of the equation by 6. Later I multiplied the right side of the same equation by 24, subtracted 4 from the result, then divided the remainder by 2 and subtracted 34 from what I have left. Have I done the same thing to both sides? No. Is the left side still equal value to the right? Check yourself. Have I done the equivalent to the right side? Yes

A Simple Equation solve for x The goal is just having x = something

$\frac{5}{8}x = \frac{21}{24}$ First eliminate the $\frac{5}{8}$ by making it become equal to 1

$$7\times3 \quad \text{Check for division first}$$

$$\frac{8}{5} \times \frac{5}{8}x = \frac{8}{5} \times \frac{21}{24} = \frac{7}{5} = 1.4 \quad \text{Again notice how helpful TOR is.}$$

$x = 1.4$

The above problem is an example of the power of understanding and how a problem, which on the surface may appear difficult, can be easily solved. It also demonstrates the versatility of What, How, and Why thinking .

Let's try a problem which is a bit more challenging.

Simplify: $\frac{5}{24} + \frac{3}{16} + \frac{5}{8} + \frac{3}{40}$

Check for factors. Yes. 8 major, no minor, individual 2, 3. *and* 5

$$\frac{5}{6\times4} + \frac{3}{4\times4} + \frac{5}{2\times4} + \frac{3}{10\times4}$$

$$3\times2\times4 \quad 2\times2\times4 \qquad\qquad 5\times2\times4$$

We can now express the Denominators with their individual factors. They will be needed for building both the LIKE SIZE which contains the common factors of each AND the LCD, Lowest Common Denominator

$$\frac{5}{3\times2\times4} + \frac{3}{2\times2\times4} + \frac{5}{2\times4} + \frac{3}{5\times2\times4} = \frac{Quotient}{LCD} = \frac{Quotient}{3\times2\times4\times2\times5}$$

We now proceed to make each Denominator become a copy of the LCD without changing its initial value.

$$\frac{2\times5}{2\times5} \times \frac{5}{3\times2\times4} + \frac{3\times5}{3\times5} \times \frac{3}{2\times2\times4} + \frac{3\times2\times5}{3\times2\times5} \times \frac{5}{2\times4} + \frac{3\times2}{3\times2} \times \frac{3}{5\times2\times4} = \frac{Quotient}{3\times2\times4\times2\times5}$$

$$\frac{10\times5}{2\times5\times3\times8} + \frac{15\times3}{3\times5\times2\times8} + \frac{30\times5}{3\times2\times5\times8} + \frac{6\times3}{3\times2\times5\times8} = \frac{50+45+150+18}{3\times8\times2\times5}$$

$$\frac{5}{24} + \frac{3}{16} + \frac{5}{8} + \frac{3}{40} = \frac{50}{240} + \frac{45}{240} + \frac{150}{240} + \frac{18}{240} = \frac{263}{240}$$

I hope you are noticing how useful and important the Three One Rules are and that you are developing the understanding and confidence to be able to confidently use them. You also need to practice making up your own problem then solve it. You have a calculator, but you only check your answer AFTER you've solved it. If you make a mistake, you know the Concepts and Rules you used, so you can turn it around, so it becomes a learning opportunity. You can always check how you're doing, one step at a time. Making up the problems yourself to solve, is very BENEFICIAL, it's increasing your understanding, self-confidence, and ability to problem-solving.

The same thinking pattern we have been using will solve any fractions problem. Don't just memorize what we did. Understand the reason for every step that was taken. MEMORIZE THE IDEAS that we used. The main reason we are doing these problems is to gain understanding, experience and practice in problem-solving. You are developing your confidence, to put ideas together in an organized, logical, scientific way. This will help prepare you for solving your future life problems. You are working with ideas. You are not just cancelling, you are eliminating problems by dividing things into themselves, making their value become equal to "1", taking advantage of TOR3, so that you can apply either TOR1 or TOR2. Learning to use Concepts, Rules, and Patterns in problem-solving is what thinking and reasoning is all about!

Let's look at something that looks a lot more complicated. You learn more by trying to do problems which you think are more difficult. If you make mistakes along the way, they are only learning opportunities. Make sure you are only making one change at a time. You'll have a much better chance of seeing what you did wrong. This will enable you to learn from your mistakes. If you change two things at the same time you won't know which change has caused the problem.

Let's see what you'll learn from the next challenge. You will simplify a problem which is made up of algebraic symbols. The problem is considered difficult by many, only because the algebraic symbols look daunting, and because there are so many of them. You'll find simplifying problems which contain algebraic symbols becomes simpler than number problems, because the symbols decrease the number of words you must keep track of. It's easier to understand the ideas you're using, and the procedure your using becomes clearer and makes sense. The main reason of feeling lack of confidence is not being able to see a familiar pattern. We will go over the ideas with numbers first, we are using very small numbers, you are familiar with them, so you will recognize the pattern easier. Let's start with 12 buttons.

12 buttons = 2×6 buttons (It helps to speak to yourself while doing the activity, 2 groups of 6); but 6 buttons are identical in value to 2×3 buttons so, 12 buttons = 2×2×3 buttons, or 3×2×2

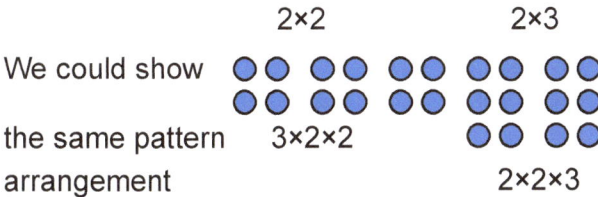

2×2   2×3

We could show

the same pattern   3×2×2

arrangement   2×2×3

I can make whatever pattern want with the 12 buttons. As long as I've used up all the buttons each time I'll still have a quantity of 12 buttons

Repeat to yourself the groupings above, but use the mathematical meaning of × "groups of" 4×3 means 3 + 3 + 3 + 3 (Four groups of three means: three + three + three + three) If I had 24 buttons I could make the following groupings:

2×12; 2×2×6; 2×2×2×3; 3×4×2; 3×8; 12×2 I could still make more groups. It's the pattern I want you to see and understand.

This time I want to use algebraic symbols. Given a, b, and c; where each symbol represents some number. When you multiply with algebraic symbols you don't place a × sign between them. It's universally understood to mean groups of (multiply in English). Let's make the following groups of abc. "abc could mean":

a groups of bc,   bc groups of a,   Any other combination of abc
b groups of ac,   ac groups of b,   would give the same value.
c groups of ab,   ab groups of c,   Assign values to a, b, and c.
a groups of cb,   cb groups of a,   2 for a, 3 for b, and 4 for c.
                                    See for yourself you always
                                    get the same value.

Given a, b, c, and d where the four algebraic symbols represent any real numbers. Given abcd: How many groups of:

a are there? _____, groups of bc _____, groups of acd _____
Ans = bcd   Ans = ad   Ans = b

groups of bd _____
Ans = ac

Given 2abcd. How many groups of 2? _____ groups of 2bc _____,
Ans = abcd   Ans = ad

groups of acd ___2b___

Groups of 2d _____ groups of abc _____, groups 2ad _____,
Ans = abc   Ans = 2d   Ans = bc

groups 2abcd _____
Ans = 1

Practice simplifying the problem on the previous page until you can, with confidence, get 100% correct regardless of the required group. You were able to do this with the number of buttons. It's much easier with the algebraic symbols. You are increasing your ability to work with Patterns, a skill required in any problem-solving situation. Below is a worked example of an addition of Fractions problem. Most of the elements are algebraic symbols

Algebra is considered difficult for many students only because of lack of understanding. When you realize you are making it easier, by using a simple symbol to decrease the amount of information you are working with, it will make more sense to use algebra.

Example problem: Simplify the following Addition of Fractions expression.

$$\frac{3}{2a} + \frac{2}{b} + \frac{4}{c} + \frac{3}{4d} = \quad \text{Where a, b, c, and d and are real numbers.}$$

We identify the factors: Minor factors: 2 in 2a, and 2 in 2×2d individual factors 2d, 2a, a, b, and c.

$$\frac{3}{2a} + \frac{2}{b} + \frac{4}{c} + \frac{3}{4d} = \frac{3}{\underset{2\times2d}{2a}} + \frac{2}{b} + \frac{4}{c} + \frac{3}{\underset{2\times2d}{4d}} = \frac{\text{Sum Numerators}}{\text{LIKE SIZE}} = \frac{\text{Sum Numerators}}{\text{a×b×c×2×2d}}$$

$$\text{product individual factors}$$

Modify, using TOR1 and 3. the denominators so that they are all LIKE SIZES

$$\frac{3}{2a} + \frac{2}{b} + \frac{4}{c} + \frac{3}{4d} = \frac{2bcd}{2bcd} \times \frac{3}{2a} + \frac{2abcd}{4acd} \times \frac{2}{b} + \frac{2ab \times 2d}{2ab \times 2d} \times \frac{4}{c} + \frac{abc}{abc} \times \frac{3}{4d}$$

$$\frac{6bcd}{4abcd} + \frac{4abcd}{4abcd} + \frac{16abd}{4abcd} + \frac{3abc}{4abcd} = \frac{6bcd + 8acd + 16abd + 3abc}{4abcd}$$

It's important that you do the above Simplify problem by yourself, analysing each step so that you know the reason, and purpose for making each step. You are not just memorizing a procedure for a test result, but developing complex, organized, logical thinking patterns that will increase understanding and self- confidence for solving more complicated everyday problem-solving

Let's look at another problem involving addition of fractions which involve variables, to see if a simpler one becomes even more simple, after the problem you've just done.

The Common Denominator we have to build is a product of the LIKE SIZES, the denominators of all the fractions we are to add, MUST be made identical to the LIKE SIZE.

Simplify the following: $\frac{1}{a} + \frac{2}{b} + \frac{3}{c} + \frac{4}{d}$ where a, b, c, and d are real numbers. You first identify the factors. There are no major or minor factors, The individual ones are a, b c, and d. The LIKE SIZE must contain all the factors, so that's: abcd. Our expression is:

$$\frac{1}{a} + \frac{2}{b} + \frac{3}{c} + \frac{4}{d} = \frac{Quotient}{LCD} = \frac{Quotient}{abcd}$$

Modifying the fractions so their denominators become equal to the LIKE SIZE

$$\frac{bcd}{bcd} \times \frac{1}{a} + \frac{acd}{acd} \times \frac{2}{b} + \frac{abd}{abd} \times \frac{3}{c} + \frac{abc}{abc} \times \frac{4}{d} = \frac{bcd + 2acd + 3abd + 4abc}{abcd}$$

Was the problem difficult? Was it easier than the Numeric One?

I have omitted most of the steps in the above simplification problem. You do the problem yourself by inserting, and justifying, the CONCEPTS, and RULES, you used. With more practice, simplification of any Addition of Fractions problem should be a straightforward task.

Following is a problem that used to be on all year 10 Australian National Assessments, and on New Zealand School Certificate National Assessments. On average, only 15% of the applicants get the correct answer. I hope you can do much better.

Simplify $\dfrac{2c}{a} + \dfrac{2c}{b}$ = I have supplied the answer below.

Simplify $\dfrac{2c}{a} + \dfrac{2c}{b} = \dfrac{2c}{a} + \dfrac{2c}{b} = \dfrac{2bc}{ab} + \dfrac{2ac}{ab} = \dfrac{2ac+2bc}{ab}$ or $\dfrac{2(ac+bc)}{ab}$

Let's Investigate Division of Fractions

This section should be a delight. You've got experience using the TOR, and now you'll see how powerful they are at simplifying division of fractions. Besides the TOR, you'll need the **Multiplication of Fraction Rules, and the Equivalents Rule.**

Equivalence Rule: Equals can always be substituted for equals.

This Rule is very important and powerful. It enables you to look at things from a different point of view. "If you want to do things better, then you have to do it differently". TOR allows you to do this.

When multiplying fractions:

You multiply the Numerators together and the Denominators together AFTER FIRST CHECKING to see if you can do any division. The division is using TOR2. You are eliminating division by making the process become equal to "1". Let's start off with a problem which seems daunting,

Simplify the following:

$$\frac{\dfrac{13}{25}}{\dfrac{39}{45}}$$ You are learning how to use What, How, and Why thinking.

$$\frac{\dfrac{13}{25} \times \dfrac{25}{13}}{\dfrac{39}{45} \times \dfrac{25}{13}} = \frac{1}{\dfrac{3}{9} \times \dfrac{5}{1}} = \frac{1}{\dfrac{5}{3}}$$

Let's try eliminating the denominator of the fraction first.

$$\frac{\dfrac{13}{25}}{\dfrac{39}{45}} \times \frac{\dfrac{45}{39}}{\dfrac{45}{39}} = \frac{13}{25} \times \frac{45}{39} = \frac{1}{5} \times \frac{9}{3} = \frac{3}{5}$$

In any complex problem involving division, it is usual that it is the division that is Making the problem difficult. No problem: TOR will always allow you to eliminate. Let's do a simpler Division of Fractions simplification. We'll still be using TOR for modifying fractions.

1. $\dfrac{8}{\dfrac{3}{4}}$

We will be using the Multiplication of Fractions Rule. We can only use this Rule efficiently if all of the quantities in the problem have their own Numerator and Denominator. We'll do this first. Using TOR2.

$$\frac{\dfrac{8}{3}}{\dfrac{4}{}} = \frac{\dfrac{8}{1}}{\dfrac{3}{4}}$$

We now have a Fraction which has a fraction as Numerator and a fraction as a Denominator we haven't changed the value of the problem. We did this using TOR2.

We'll simplify the Fraction by getting rid of the Denominator. We can do this with TOR3

$$\frac{\dfrac{8}{3}}{\dfrac{4}{}} = \frac{\dfrac{8}{1}}{\dfrac{3}{4}} \times \frac{\dfrac{4}{3}}{\dfrac{4}{3}} = \frac{\dfrac{8}{1} \times \dfrac{4}{3}}{1} = \frac{\dfrac{32}{3}}{1} = \frac{32}{3}$$

We have maintained the integrity of the problem because we have only multiplied the whole problem by 1. We can always get rid of division with a combination of TOR2 and TOR3. TOR2: Anything divided by "1" remains the same in value. TOR3: Anything divided by itself, excluding divide by zero is "1"

You have very neatly eliminated each obstacle with a rule, and you are using patterns that have been successfully used in previous problems. For visual clarity I have not used strike outs to show which elements have been **eliminated** by being divided by themselves. With patience you can see what has happened and these are very important skills to learn. It's important that you understand why each step was done, and that, in each case, it eliminated the obstacle.

You will be using the above steps in any division of fractions situation. Through understanding the reason and purpose Why you are using the Rules and Concepts you are, they become imbedded in your thinking process. By using them in problem solving you become more confident, and capable of using these tools. Remember using mental tools with understanding and confidence is how you become a better problem-solver in all the problem-solving activities you will be involved later in life.

Example Simplifying of fractions:

This is only a slight modification of what we have already done.

3. Given $\dfrac{\frac{5}{12}}{\frac{15}{16}}$

$$\dfrac{\frac{5}{12}}{\frac{15}{16}} = \dfrac{\frac{16}{15}}{\frac{16}{15}} \times \dfrac{\frac{5}{12}}{\frac{15}{16}} = \dfrac{\frac{16}{15} \times \frac{5}{12}}{1} = \dfrac{4 \times 4 \times 5}{3 \times 5 \times 3 \times 4} = \dfrac{4}{3 \times 3} = \dfrac{4}{9}$$

Be sure to softly say to yourself, why you are using the Rule or Concept you used each time. Doing this imbeds the process you have used in your thinking patterns, which you will incorporate by yourself, into other problem—solving activities I will not always put in all of the steps. As you gain more understanding you will be able to do this yourself.

Once more the Rule for Multiplication of Fractions

Multiply the Numerators together AFTER FIRST CHECKING to see if there is Division.

An example of Simplifying a fraction problem:

4. Simplify

$$\frac{\dfrac{4a}{15bc}}{\dfrac{2}{25cd}} = \frac{\dfrac{25cd}{2}}{\dfrac{25cd}{2}} \times \frac{\dfrac{4a}{15bc}}{\dfrac{2}{25cd}} = \frac{\dfrac{25cd}{2} \times \dfrac{4a}{15bc}}{1} = \frac{5d}{1} \times \frac{2a}{3b} = \frac{10ad}{3b}$$

$$\frac{\dfrac{4a}{15bc}}{\dfrac{2}{25cd}} = \frac{10ad}{3b}$$

Let's take a look at Equations

An Equation is used when it is necessary to communicate your ideas in a Problem-solving, Logical, Scientific manner, that requires:

- Clarity
- Correctness and
- Conciseness

When problem solving intellectual tools are used. These include:

- Concepts
- Rules
- Patterns
- Shapes
- Anything which influences your thinking, feeling, or behaviour

The language used in Mathematics is very concise:

- One symbol, only one meaning
- One word, only one meaning
- One sentence, only one meaning

It's the conciseness and correctness that allows ideas to be logical. You say what you mean and mean what you say. There is no reading between the lines. One idea naturally follows from the previous idea.

A Concept is the shortest string of words, that conveys understanding and meaning. It always tells you what to do. It's a mental tool.

A Rule is the shortest string of words, that conveys, understanding and meaning. It always tells you how to do it. It's a mental tool.

A pattern contains a string of meanings that when connected, explain an event, a thought, an idea, or an instruction. The pattern is reproducible and can be modified to explain a new event.

When using equations, you will be using a mixture of algebraic symbols and algebraic variables. It's helpful to know when symbols need to be defined and when they are universal in use and therefore already understood. The General Equation is always understood by all who use that equation. It comes under the classification of a mental tool; it is a proven Rule. $F=ma$ is the general equation for force where m and a stand for mass multiplied by acceleration. $V = 6$ ml would stand for a quantity of 6 millilitres of liquid or gas. $\perp$ is the General Symbol for "perpendicular". + is General symbol for "Make groups of Like Things". When multiplying variables, it's accepted internationally, that the multiplication symbol is not required between them. When using addition, subtraction, or division, correct symbols (Upper or Lower case) are required.

When working in algebra, for clearness, correctness, and simplicity you are using simple symbols that stand for a single word, sentence, or whole paragraph. There's no magic or complexity in that, only remembering what the symbols stand for. They convey information and understanding Clearly, Correctly and Concisely (the three c's). Only the Mathematics language can do this.

The algebraic variable can be used to define whatever you're working with. It clarifies and helps make ideas easier to understand and to work with. For example: $v = d/t$ stands for "velocity is equal to the distance travelled, divided by the time taken to travel that distance".

When solving problems you use tools. Either material tools or mental tools. The equation uses the mental tools called, Concepts, and Rules; The equation is the Powerhouse of problem-solving. It allows simple, reliable, logical, efficient manipulation of information or ideas. If you're using an equation and are manipulating your ideas logically and using the correct Concepts and Rules, you can be confident that your answer is correct. It is only by using the equation, that we have been able to design and build all the modern equipment and conveniences that we have today.

Your brain is solving problems 24/7 even when you're asleep. When you wake, you share a common problem with everyone else, the battle between what you'd like to do, and what you should do. When the problem involves a brand-new task, but you've not been told exactly what to do, and how to do it, can you correctly start? The obvious answer is NO! You don't know what mental or material tools you need, and how to begin. Every logical problem starts the same way. It starts with knowing:

- **What to do**.
- **How to do it. and**
- **Why you're doing it,**

When it's working with people the **Why** encompasses: "Why are you thinking, feeling, or behaving this way", and those ideas should be based on the virtues. There must be a good reason for your behaviour, and they should be based on honesty, respect, kindness, trustworthiness, thoughtfulness etc. If it isn't a good reason. Why are you doing it?

If you had a problem that required material tools, you'd start collecting the material tools you need to complete the task. If you lacked the understanding, knowledge, and experience to do the job, and didn't have the right tools, you'd have great difficulty completing it.

When the problem requires good thinking, it follows the same formula, but the formula should be based on: What, How, and Why, are you doing it? It's always the Why which keeps you on the right track. You are learning to Reflect, Visualize, and use Patterns. It's important to realize: **It's not how much effort you put into a task, it's effort in the right direction which enables you to complete it.**

Before and after each step in problem-solving, you reflect and visualize. Why you are Thinking, Feeling, and Behaving the way you are? Using the **Why** must become a habit you are learning to Reflect and visualize (look before you leap). You're Verifying the correctness of each step, by checking your result:

- Why are you going to use that Concept?
- What did you expect to achieve?
- Did you achieve what you expected?
- If not, why not.? You have only made one step forward so you should be able to correct the choice.
- If Yes, What Rule are you going to use?
- What did you expect to achieve?
- Did you achieve what you expected?
- If not, why not? You have only made one step forward so you should be able to correct the choice.

- A mistake is not a mistake, but an excellent learning opportunity. You used your present understanding, knowledge, experience and wisdom, yet it didn't work. Why didn't it work? If you didn't make a mistake, you're happy, and you've gained more confidence, but you've only reinforced what you already knew. It's by correcting, and figuring out what caused a mistake, and learning from it, that you grow.
- Realize, the effort made, and what you gained from making the effort, will increase your understanding, resilience, and confidence towards becoming a good problem-solver.

Knowing how to problem-solve and having enough self-confidence are very important for everyday problem-solving. You need some sort of blueprint which teaches you, or shows you, how to put ideas together in a meaningful logical way for problem-solving.

Many of our everyday problems involve dealing with people and communities. In this situation the **Why** must also include the virtues. The virtues are thinking, honesty, kindness, thoughtfulness, patience trustworthiness, hard-working, resourcefulness etc. There're over 100 virtues. They're the basic life skills in every religion. Remember we are all capable of thinking and analysing, but at a different rate. Be patient. Give the other person the time to understand what you're communicating. Realize, when you're using English to communicate an idea, or an instruction, it's not like the language of Mathematics. In English a symbol can have several different names. A word can have eight or more different meanings. If we had a short five-word sentence and we averaged out each single word of the five words to have five different meanings, the number of possible combinations of meanings is mathematically $5 \times 5 \times 5 \times 5 \times 5$ that's over 3,000 possibilities. Fortunately, the brain is going to look for familiar word patterns, frequently used words, and a number of other short-cuts, and of course the brain works very quickly. The point I'm making is that it is very difficult to communicate very discrete meanings when there are so many possibilities for misunderstanding.

This is where Mathematics excels. It's a subject, which, **when taught differently**, can create caring, thoughtful, honest, confident, efficient, caring, persevering, resourceful, dependable, hardworking (The Virtues) people. When working with Mathematics you are working with ideas. Working with numbers is only a very small part of what you are learning. That's why Mathematical thinking is so important. It's all about learning how to think more efficiently, and that's important in any situation.

When using a Mathematics equation, you are always goal orientated:

- You always have all the mental tools you need to handle the task.
- You are solving problems by using Concepts, Rules, and Logical thinking.
- Concepts and Rules are facts.
- Concepts and Rules always: tell you What to do, and How to do it.
- You are eliminating the obstacles in the way of the goal, one by one, one step at a time.
- Remember a mistake is not just a mistake it's an excellent learning opportunity.
- The effort made, and what you gain from making the effort, will increase your understanding, resilience, and confidence towards becoming a good problem-solver.
- If you didn't make a mistake, you're happy, and you've gained more confidence, but you've only reinforced what you already knew. It's by correcting, and figuring out what caused a mistake, and learning from it, that you grow.

By following the above procedure, you are gaining, understanding, self-confidence, problem-solving skills, as well as developing the higher-level thinking skills you need for the more difficult tasks.

Summarising: It's very important, when using equations, only Mathematics words, meanings, and terms are used. "Remember: one symbol, one word, one sentence; they all have only one meaning". There is no confusion in meaning or understanding. The strength of the Equation is that it enables you to be goal orientated; Able to logically, and with understanding and self-confidence, efficiently, manipulate facts; Able to derive new ideas, Concepts, and Rules that are confidently understood.

Equation Concept – The left side of the equals symbol has the identical value, or meaning, as that on the right side, or vice versa.

Equation Rule – You can do whatever you like to one side of the equation, **providing**, you do the equivalent to the other side. Equivalent in mathematics means identical in value or meaning.

Herein lies the power "always goal orientated, providing you do the equivalent to the other side".

Many Essential Mental tools needed for manipulating Equations are listed below.

You were introduced to the **Three One Rules** in Chapter 6. You'll see how powerful, and useful they are for manipulating fractions and equations. Working with equations means understanding fractions so that you are confident in their correct and proper use during problem-solving.

Fraction Concept – A fraction is a division instruction.

A Fraction has the following format: were x and y stand for quantities or things, and the line between them means divide the top by the bottom.

The top (x) above the fraction line is called the numerator. The bottom (y) below the division line is called the denominator.

Fraction Rule – Divide the Numerator by the Denominator.

Multiplication Rule for Fractions - Multiply **Numerators** together, and the **Denominators** together after first checking to see if you can divide.

Equivalence Rule – Equivalents can always be substituted for Equivalents. They are always identical in value. For example: 2×3 is equivalent in value to 6; They are not the same as 6. If you were working in a shop and the customer wanted 2, 3 kg bags of flour. The customer might not be happy with 1×6kg bag of flour. He would be less happy if you tried to say that they are equal. He needed one bag for the next-door neighbour and one bag for himself. 3×4 is identical in value to 4×3 but is not the same as 4×3; two fifty cent pieces are identical in value to a $1 coin, but if the vending machine only takes $1 coins you will be in trouble if you only have 50 cent pieces.

Let's use as an example the following simplify problem:

Example 1 $\quad \dfrac{3}{4} \times 4 \times \dfrac{3}{16} \times 8 \times \dfrac{3}{20}$

You are multiplying fractions so all of elements must be Fractions. This can easily be done by using TOR2. (Anything divided by One remains the same in value). The equation now becomes

$$\dfrac{3}{4} \times \dfrac{4}{1} \times \dfrac{3}{16} \times \dfrac{8}{1} \times \dfrac{3}{20}$$

You can now use the Multiplication of Fractions Rule, first you check for division.

You notice you can do division if you use the Equals Rule: 2×8 can be substituted for the 16 in the denominator allowing you to use TOR3 (Anything divided by itself is one excluding dividing by zero) allowing you to divide the 8 in the numerator by the 8 in the denominator.

You can substitute 5×4 for 20 in the denominator this allows you to divide the 4 in the numerator by the 4 in the denominator.

You then notice you can do the same thing substituting 2×2 for the 4 in the numerator, which allows you to divide out the twos.

You also eliminated the multiplication by 1 using TOR1.

The process just described is demonstrated below: Each step was a logical step and followed a definite Rule. With practice you could do the steps which were done in your head.

$$\frac{3}{4} \times \frac{4}{1} \times \frac{3}{16} \times \frac{8}{1} \times \frac{3}{20} \ \Rightarrow\ \frac{3}{4} \times \frac{1}{1} \times \frac{3}{2} \times \frac{1}{1} \times \frac{3}{5} \ \Rightarrow\ \frac{3}{4} \times \frac{3}{2} \times \frac{3}{5}$$

$$2 \times 8 \qquad 5 \times 4$$

There is no other division so you can continue with the multiplication.

Your final simplification gives you: $\dfrac{27}{40}$

Example 2

The Three One Rules: They are used left to right or right to left.

1. Anything multiplied by One remains the Same in value. Anything remains the same in value if it is multiplied by One.

2. Anything divided by One remains the same in value. Anything stays the same in value if it is divided by One.

3. Anything divided by itself is One (Exclude divide by zero. Its equivalent to Asking: "What's the largest number").

For example: What happens to the size of the Numerator as the Denominator gets smaller and smaller approaching zero. Would it ever reach zero. No, the Denominator would continue to get smaller, and Numerator would just continue getting larger. The symbol we use for this is the infinity symbol " ∞ ". If you were using a computer you would have to insert in your program a check, having it stop calculating when the necessary accuracy was reached.

$$\frac{1}{1} = 1; \quad \frac{1}{0.1} = 10; \quad \frac{1}{0.01} = 100; \quad \frac{1}{0.001} = 1000; \quad \frac{1}{0.0001} = 10,000$$

Equivalent Rule – Equivalents can be exchanged for equivalents. Equivalent is defined as equivalent in value or meaning. Equals is not a mathematical expression "It has many meanings and therefore doesn't fit the bill for clear, accurate, and concise. It creates confusion and misunderstanding". You would use the phrase "identical in ……. value, colour, size, height etc.

Manipulating Equations.

F = *ma*. Where F = Force, *m* = mass, and *a* = acceleration.

The above equation is the General Equation for Force. As long as you know the General Equation you can easily manipulate the terms to find the equation for mass or acceleration. Let's determine the equation for acceleration using the General Equation for Force. Starting with:

$F = ma$ We would like the equation to be: m = something. That means there is only m on one side of the equation. We would have to rearrange $F = ma$ so that "a" (acceleration) becomes equal to one without changing the meaning of the expression . No problem, we need to use TOR3. We will divide a by itself. When we have an equation we can do whatever we like to one side, providing we do the equivalent to the other side. First we make F and ma into fractions.

$$\frac{F}{1} = \frac{ma}{1}$$ TOR1 has not changed the value of anything on one side, so nothing has to be changed on the other side,

$$\frac{1}{a} \times \frac{F}{1} = \frac{ma}{1} \times \frac{1}{a}$$ We will get the equation m = something by multiplying both sides of the equation

$$\frac{F}{a} = m$$

$$\frac{F}{m} = a$$

by $\frac{1}{a}$ We could do the same procedure, except dividing by m, if we wanted the equation of a = something,

Let's make the equation a bit more complicated. Given the imaginary equation: let's make the equation a = something

$$\frac{3}{4} \times \frac{abcd}{1} = \frac{5}{1} \times \frac{ad}{c} \times \frac{4}{3a^2cd^3}$$ This simplifies to $$\frac{3abcd}{4} = \frac{5ad}{c} \times \frac{4}{3a^2cd^3}$$

Determine the value of b in terms of the variables, $a, c,$ and d. Our goal is: b = something.

Starting with $$\frac{3abcd}{4} = \frac{5ad}{c} \times \frac{4}{3a^2cd^3}$$ We want to have only b on the left.

We proceed to make everything other than b become equal to 1 and do the equivalent to what we did to the left side to the right side.

$$\frac{4}{3acd} \times \frac{3abcd}{4} = \frac{4}{3acd} \times \frac{5ad}{c} \times \frac{4}{3a^2cd^3}$$ We can now simplify.

We notice that there is division we can do, so we do that first.

$$b = \frac{4}{3acd} \times \frac{5ad}{c} \times \frac{4}{3a^2cd^3} = \frac{4 \times 5 \times 4}{3c \times c \times 3a^2cd^3} \times \frac{80}{9a^2c^3d^3}$$

$$b = \frac{80}{9a^2c^3d^3}$$

Notice that each time, before I make a move, I'm reflecting and visualizing that I'm using the correct Rule or Concept. You are trying to make this a habit. It will soon become automatic. You are building self-confidence in that, you know what you're doing, how you're going to do it, and why you're doing it (each move is going towards your goal). You are practicing making decisions based on cold hard facts. You are learning to think logically. You can depend on your conclusion. The above pattern of thinking is very important. You don't need to memorize a lot of equations. You only need to know how to manipulate an equation to get what you want.

The skill you're trying to learn and develop is the ability to confidently put your ideas together in the form of an equation. Once you have the equation you are in a situation of POWER. That is the only time when you can confidently feel you oversee things. Make sure you are learning the General Equations. Later you'll find you're substituting a General Equation into a General Equation.

The previous problem required some knowledge of working with Indices. It would be a good idea to refresh your knowledge as we will be using more sophisticated ideas later in the chapter. We took advantage of a pattern to define the Index laws. For example:

Using x to represent any real number we have:

x to be identical in value to x^1
xx to be identical in value to x^2

xxx to be identical in value to x^3

xxxx to be identical in value to x^4 and so on

It was easy to extend this pattern to the following situation

$x^2 x^3$ means <u>xx</u> groups of *xxx* but that's identical to *xxxxx* which is identical in value to $x^{(2+3)} = x^5$ You are following a logical pattern. You can depend on it. You then can use one of the Index Rules: When you have Like Things multiplied together which have indices you add the indices together:=>

(1) $x^1 x^2 x^3 = x^6$ (2) $x^{(2+3)} \times x^6 = x^{12}$ (3) $a^{-3}b^2a^4b^{-1} = ab$

(4) $2a^2b(3a^3b^2 + 4ab + 6\,a^4b^3) = 6a^5b^3 + 8a^3b^2 + 12a^6b^4$

Wait a minute! What would x^0 be equal to? If we are to investigate what x^0 means we need an equation with this unknown value. No problem. The Equation Concept will allow this.

We start with $x^0 = x^0$ and we keep the left side constant and only manipulate the right-hand side. Whatever you end up with on the right side MUST have the identical value as the left. The Equals Concept tell us this.

We start with $x^0 = x^0$

We are starting with an index which has a value that we are uncertain of. If we can use an equation to determine its value, and we follow all of our mental tools correctly we can depend on our result. We are going to use TOR1; we must multiply with a value which has an index. We'll use x and y where x and y are any real numbers

We start with:

$x^0 = x^0$ and y is any real number

$$x^0 = \frac{x^0}{1} \times \frac{x^y}{x^y} = \frac{x^{0+y}}{x^y} = \frac{x^y}{x^y} = 1$$

This demonstrates that any real number to the index 0 = 1

We have followed TOR1 and TOR2 and the equation Concept and Rule. The left side must be identical in value to the right side. Our conclusion is that for all real numbers x^0 must be equal to 1. Mathematics is a Problem-solving, Logical, Scientific language and we followed the language explicitly. So, what -ever we ended up with must be identical in value to the left side.

You are gaining experience working and thinking with Concepts and Rules, as well as building confidence and thinking logically; one idea naturally following the previous idea; and gaining confidence in the reliability of your answer.

Let's investigate another situation. What does x^{-1} mean? Our previous method was very successful and makes sense. Having what we are looking for on the left, keeping its value constant, and manipulating the right-hand side to see what result we achieve.

$$x^{-1} = x^{-1} \times \frac{x^{+1}}{x^{+1}} = \frac{x^{-1}}{1} \times \frac{x^{+1}}{x^{+1}} = \frac{x^0}{x^{+1}} = \frac{1}{x^{+1}}$$

Thus, we can say, x^{-1} is always equal to its reciprocal with a positive index.

What about $x^{1/2}$ what does that mean. First we'll try the same previous successful technique.

$$x^{1/2} = x^{1/2} \times \frac{x^{-1/2}}{x^{-1/2}} = \frac{x^0}{x^{-1/2}} = \frac{1}{x^{-1/2}}$$

This hasn't helped, we already knew this, we'll have to have a new think!

It's the fractional index that's causing us trouble. Let's use the same pattern, we set what we want equal to an equivalent value.

We want to find the value of $x^{1/2}$. We'll follow the same type of pattern as before, but look at if from a different point of view.

What do we know about $x^{1/2}$.

$$x^{1/2} \times x^{1/2} = x^{(1/2+1/2)} = x^1 = x$$

This looks very promising, let's try the following:

Let $\quad x^{1/2} = y \quad$ Where y is a real number

$$(x^{1/2})^2 = y^2$$
$$(x^{1/2})^2 = x^1 = x = y^2$$

This means x is equal to some number multiplied by itself.

Thus: $\qquad 9^{1/2} = 3 \qquad$ What number multiplied by itself is 9

$\qquad\qquad 25^{1/2} = 5 \qquad$ What nmber multiplied by itself is 25

$\qquad\qquad x^{1/2} = 36 \qquad$ What number multiplied by itself is 36

$\qquad\qquad\qquad\qquad$ x must be equal to 6

This same pattern would continue for:

$\qquad\qquad x^{1/3} = $ What number multiplied by itself three times

$\qquad\qquad x^{1/4} = $ What number multiplied by itself four times

$\qquad\qquad 27^{1/3} = 3$

$\qquad\qquad 64^{1/4} = 4$

A Table of the results of the above is given below:

If $x^{1/2}$ Is equal to:	Then x equals:	If $x^{1/2}$ is equal to:	Then x equals:	If $x^{1/2}$ is equal to:	Then x equals:
1	1	100	10	10000	100
4	2	400	20	40000	200
9	3	900	30	90000	300
16	4	1600	40	160000	400
25	5	2500	50	250000	500
36	6	3600	60	360000	600
49	7	4900	70	490000	700
64	8	6400	80	640000	800
81	9	8100	90	810000	900

The Table above demonstrates a reproducible pattern, robust and transferable

Index 10 Table for an index of Ten multiplied by itself

Hundreds	Tens	Ones	Tenths	Hundredths
100×100 = 10,000	10×10 = 100	1×1 = 1	0.1×0.1 = 0.01	0.01×0.01 = 0.0001
200×200 = 40,000	20×20 = 400	2×2 = 4	0.2×0.2 = 0.04	0.02×0.02 = 0.0004
300×300 = 90,000	30×30 = 900	3×3 = 9	0.3×0.3 = 0.09	0.03×0.03 = 0.0009
400×400 = 160,000	40×40 = 1600	4×4 = 16	0.4×0.4 = 0.16	0.04×0.04 = 0.0016
500×500 = 250,000	50×50 = 2500	5×5 = 25	0.5×0.5 = 0.25	0.05×0.05 = 0.0025
600×600 = 360,000	60×60 = 3600	6×6 = 36	0.6×0.6 = 0.36	0.06×0.06 = 0.0036
700×700 = 490,000	70×70 = 4900	7×7 = 49	0.7×0.7 = 0.49	0.07×0.07 = 0.0049
800×800 = 640,000	80×80 = 6400	8×8 = 64	0.8×0.8 = 0.64	0.08×0.08 = 0.0064
900×900 = 810,000	90×90 = 8100	9×9 = 81	0.9×0.9 = 0.81	0.09×0.09 = 0.0081

The pattern is the same as when we did the division of integers, BUT THERE IS A BIG DIFFERENCE. To satisfy all the values for 1-9 you need two sizes: 1 – 81. To satisfy all the sizes for 10 – 90 you need four sizes: 100 – 8100. To satisfy all the sizes for 1000 – 9000 you need six sizes: 10000 – 810000. The same pattern occurs for the values less than one. Except they are two smaller sizes each time.

If we use our Power of Ten notation and the Index Rule, it is much easier to follow. For Example:

$1×10^0 × 1×10^0 = 1×10^{0+0} = 1$ Which is one size

$1×10^1×1×10^1 = 1×10^{1+1} = 1×10^2 = 100$ Which is two sizes

$1×10^2×1×10^2 = 1×10^{2+2} = 1×10^4 = 10000$ Which is four sizes

$1×10^3×1×10^3 = 1×10^{3+3} = 1×10^6 = 1000000$ Which is six sizes

$1 \times 10^{-1} \times 1 \times 10^{-1} = 1 \times 10^{-(1+1)} = 1 \times 10^{-2} = .01$ Which is two sizes to the left.

The pattern repeats itself. The fractional indices are called roots.

$9^{\frac{1}{2}}$ would mean the square root of 9; what number multiplied by itself is 9. In this case we know from our table the square root of 9, it's 3 The Maths *symbol for square root is* $\sqrt{}$

$27^{\frac{1}{3}}$ Means, what is the third root of 27; what number multiplied by itself three times; $3 \times 3 \times 3$

$81^{\frac{1}{4}}$ Means, what is the fourth root of 27; what number multiplied by itself four times; $3 \times 3 \times 3 \times 3$

The pattern continues indefinitely. To solve square roots, we will use a pattern similar to the one used for division. There is a major modification to the pattern, you are now working with a number multiplied by itself. In the process of Division, we made up the table of the single digits' multiplication up to 9.

Working with fractional indices I have made up a Table of digits multiplied by itself. You require working with algebraic variables and second degree, quadratic equations. Equations of the form $x^2 + 2xy + y^2$ which had been planned for Chapter 10. Unfortunately, that will have to be included in a second book.

Conversions and Equivalents

Basic Common Relationships

Prefixes:
Below are the fractions or multiples to be used against the Standard Units'
For example:

Decimetre means a length identical to 1/10 the Standard Unit of one metre, kilometre means a length identical to 1,000

Standard Unit of one metre Gigametre means a length identical to 1,000,000,000 Standard Unit of one metre

Giga = 1,000,000,00	Deci = 1/10
Mega = 1,000,000	centi = 1/100
kilo = 1,000	milli = 1/1000
hector = 100	micro = 1/1000/000
deka = 10	nano = 1/1,000,000,000
	pico = 1/1000,000.000,000

Standard Unit of length: metre; m
Standard Unit of volume: litre; l
Standard Unit of mass: gram; g
Standard Unit of Area: m^2; cm^2; km^2
Standard Units of Time: year, yr; month, mon; week, wk; day, dy; hour, hr minute, min; second, sec

Time equivalents: 60 sec = 1 min;
                     60 min = 1 hr;
                      24 hr = 1 dy;
                       7 dy = 1 wk
                      52 wk = 1 yr
   For accuracy 365.25 days = 1 yr

Volume to Linear Equivalents: $1 cm^3$ = 1 ml
Metric to Imperial measurements: 1 in = 2.54 cm (in = inch)
Imperial Units Equivalents: 12 in = 1 ft. ft = foot
Basic Time: year, month, week, day, hour, minute, second

For Conversion and Equivalence problems it is only necessary to memorize the basic units of measure. Using Reflection and Visualization you realize that, in these types of problems, or any others, "IF YOU ARE NOT CHANGING THE VALUE OF WHAT YOU HAVE, ONLY EXCHANGING ONE UNIT OF MEASURE FOR ANOTHER, YOU ARE ONLY MULTIPLYING THE UNIT BY AND EQUIVALENT "ONE"". If you know you are only multiplying or dividing by itself, you are using TOR. If you know the Standard Unit of measure, and the equivalent, you can easily convert one value in one unit for an identical value in another unit. These problems are ideally suited for developing Higher Level Thinking and Reasoning using the intellectual Tools shown below:

- Three One Rules,
- Fraction Rules, and
- Equation Rules.

With these mental tools you can calculate and manipulate back and forth easily and efficiently. The more problems you do which require putting ideas together in an organized logical manner, the higher level of thinking you're doing. You're also gaining logical problem-solving experience, and self- confidence.

With the use of the Three One Rules you can:

- Multiply by "1" without changing the value.
- Divide by "1" without changing the value.
- Get the suitable "1" you need by dividing an equivalent by an equivalent.
- With simple organized logical steps reach the desired unit of measure.

Following are examples of the unlimited combinations of "Ones" you can make up.

$$Examples\ of\ \frac{Equivalent}{Equivalent} = \frac{60\ sec}{1\ min} = \frac{1\ min}{60\ sec} = \frac{7\ dy}{1\ wk} = \frac{12\ months}{1\ year} =$$

$$\frac{\$1}{100\ cents} = \frac{1000g}{1kg} = \frac{1yr}{365.25\ dy} = \frac{5{\times}20\ cent\ pieces}{\$1} = etc.$$

Examples of Solved Conversion problems are shown below:

To change the following into the required unit. You always start with what you've got, then convert using the basic units, to what you want.

Example 1.

Change 4 meters into centimeters

$$4m = \frac{4m}{1} \times \frac{100cm}{1m} = \frac{400cm}{1} = 400cm$$

First you made 4m into a fraction by using #2 Rule of the Three One Rules.

Second you have made up the appropriate "1"

You needed to eliminate the unit "m", You could do this by dividing the m by itself you needed an appropriate fraction (an instruction to divide).

You created $\dfrac{100cm}{1m}$ following the Fraction Concept and Rule.

You then proceded to carry out the instructions. Your steps were clear, correct, concise, and made sense.

Example 2.

Given 3 days, change into an equivalent number of seconds

$$3dys = \frac{3dy}{1} \times \frac{24hr}{1\,dy} \times \frac{60min}{hr} \times \frac{60sec}{min} \times \frac{3\times24\times60\times60sec}{1} = 259,200sec$$

In the above procedure I have put in every step for clarity. As long as you are continuing to reflect and visualize and understand WHY YOU ARE DOING WHAT YOU ARE DOING it's not necessary to show every new position that has occurred.

Example 3.

Given 3*ft* convert to *cm*

$$3ft = \frac{3ft}{1} \times \frac{12in}{ft} \times \frac{2.54cm}{in} = \frac{3\times12\times2.54cm}{1} = 91.44cm$$

Example 4.

Given the following Equivalents: 1 yard (yd)=3 feet (ft) 12 in =1 ft, 2.54 cm = 1 in

Convert 1 yd into cm.

$$\frac{1yd}{1} = \frac{3ft}{1yd} \times \frac{12in}{1ft} \times \frac{2.54cm}{1in} = \frac{3\times12\times2.54cm}{1} = 91.44cm$$

Example 5:

Given 45 ft³ Convert into m³.

$$\frac{45ft^3}{1} = \frac{45ft^3}{1} \times \frac{12\times12\times12in^3}{ft^3} \times \frac{2.54\times2.54\times2.54cm^3}{in^3} \times \frac{m^3}{10^6cm^3}$$

$$= \frac{45\times12\times12\times12\times2.54\times2.54\times2.54m^3}{10^6} \times \frac{10^{-6}}{10^{-6}}$$

$$45ft^3 = \frac{1274258.1m^3}{10^6} \times \frac{10^{-6}}{10^{-6}} = \frac{1274258.1m^3\times10^{-6}}{10^0} = 1.27m^3$$

Example: This example conversion is important for all future and present drivers. Convert. 30 km/hr into m/sec

$$\frac{30\ km}{hr} \times \frac{1000m}{km} \times \frac{hr}{60min} \times \frac{min}{60sec} = \frac{10^3m}{2\times60sec} = \frac{10^3m}{12\times10^1sec} = \frac{8.3m}{sec}$$

Driving in town is usually 60km/hr that's 16.6m/sec that's almost a blink

$$\frac{80km}{hr} = \frac{80km}{hr} \times \frac{hr}{60min} \times \frac{min}{60sec} \times \frac{1000m}{km} = \frac{80 \times 10^3 m}{36 \times 10^2 \times sec} = \frac{22.2m}{sec}$$

$$\frac{100km}{hr} = \frac{100km}{hr} \times \frac{hr}{60min} \times \frac{min}{60sec} \times \frac{1000m}{km} \times \frac{100 \times 10^3 m}{36 \times 10^2 \times sec} \times \frac{28.2m}{sec}$$

A little over 28 m is a long way to travel in just a blink. Looking at your cell phone while driving would take a few blinks. This is why looking at the telephone even at 60 km/hr is extremely dangerous!

The above ideas are what you use in any conversion situation. Go over all the steps again. You are practicing What, How, and Why, thinking and reasoning. Being able to do conversion problems is something you need to learn for outside the classroom. You just can't push buttons on the calculator and expect to have a reliable answer. It's necessary to also understand what you're doing.

Percent. Comparing and Measuring things by using a common size Denominator.

I said we would cover Percent later in the Chapter so here we go. Why do we need Percent when we already are able to compare quantities by using a fraction? Well, it's easy to compare when the fractions are ½ , ¾, 1/5, or another other familiar sizes, but what happens when the fraction is not familiar? in fact the fraction could be any combination of numerator and denominator comparing parts of a whole. It's very hard to quickly compare quantities of an article when you are comparing two different fractions of the same element and need to know which is the larger or smaller one. For example: In the statement below, which is larger a) or b)?

a) 147/350 means 147kg rice in a 350 kg box or b) 416 kg rice in an 850 kg box If you had to purchase a) or b) and the price per kg was the same for each box which is the better buy? This type of situation is an everyday situation for people buying and selling goods. Which one do you think is bigger? Below are the fractions presented as Percents.

a). 42 percent of the box are rice and 49.1 percent for b). These sort of transactions, involving many Billions of dollars, are taking place every day. With large purchases a single percent could mean millions of dollars. It is this reason that percent is the figure which is used for comparing two or more things. The good news expressing things as a percent is very easy. First you need to know what percent means.

When you say ½ the bag is empty your x/y value is mathematically equal to a denominator which is exactly twice as large as the numerator, but when that isn't true it becomes harder to know the value of x/y. It would become easier if you had a common size denominator in every transaction. Universally that size denominator is 100. It can easily be calculated if it was 3/5, in you head you would have said there are 20×5 in 100 and 20×3 is 60 Percent. You were expressing the quantity as the number of 1/100. Without realizing it you were using TOR the Three One Rules.

In the first instance you mentally did the following:

$$\frac{3}{5} \times \frac{20}{20} = \frac{60}{100} = 60 \times \frac{1}{100} = 60\%$$

The percent symbol % means "Number of 1/100

But what do you do when the fraction is $\frac{9}{13}$? That's a job for TOR. You want to make the 13 become 100 without changing the value of the fraction. No problem. Make the 13 become 100 first them make sure you're only multiplying the whole problem by 1.

$$\frac{9}{13} \times \frac{100}{\frac{100}{13}} \Rightarrow \frac{\frac{9}{1}}{\frac{13}{1}} \times \frac{\frac{100}{13}}{\frac{100}{13}} = \frac{9}{1} \times \frac{100}{13} = \frac{9 \times 100}{13} = \frac{9}{13} \times 100$$

Eureka!

We have come up with a helpful new idea. It looks like all you must do to change a fraction to the number of 1/100 is multiply the fraction by 100. We'll have to check our idea out with algebra to make sure before we can call it a rule.

To prove multiplying by 100 will change the fraction to percent we'll do the following: Let x and y be any real number. Then check to see what happens when any fraction, or number is multiplied by 100. We'll **start by setting what we want equal to itself.** Its very important that you go over this proof carefully thinking out loud why each step was done.

This sort of thinking can be attached to other situations and can become part of your thinking strategies. The more you use this type of thinking the more you will be able to incorporate the thinking pattern into your normal problem-solving procedures you are already using. It will increase your self-confidence which will allow you to become an even better problem-solver.

$$\frac{x}{y} \times 100 = \frac{x}{y} \times 100$$

$$\frac{x}{y} \times 100 = \frac{x}{y} \times 100 = \frac{\frac{x}{1}}{\frac{y}{1}} \times \frac{\frac{100}{y}}{\frac{100}{y}} = \frac{\frac{x}{y} \times \frac{100}{1}}{100} = \frac{\frac{x}{y} \times 100}{1} = \frac{x}{y} \times 100 \times \frac{1}{100}$$

This demonstrates that for any $\frac{x}{y}$ multiplying by 100 will give the quantity $\frac{x}{y} \times 100$ hundredths $= \frac{x}{y} \times 100$ percent

But the number of 1/100 is percent which allows you to write your answer as $\frac{x}{y} \times 100\%$

Solve the following for y. You can always check your answer by substituting the answer from the solution back into the original equation. More examples of using the Three Rules follow: We are always rearranging the equation so that we only have "y" the variable on one side. We aim to have the variables coefficient become equal to 1.

1. $\quad \dfrac{3y}{4} = \dfrac{5}{7}$ *we aim for* y = *something* $\qquad$ **2.** $\quad \dfrac{3}{7}y = \dfrac{5}{3}$

$\dfrac{4}{3} \times \dfrac{3y}{4} = \dfrac{5}{7} \times \dfrac{4}{3} \qquad\qquad\qquad \dfrac{7}{3} \times \dfrac{3}{7}y = \dfrac{5}{3} \times \dfrac{7}{3}$

$\qquad\qquad y = \dfrac{20}{21} \qquad\qquad\qquad\qquad\qquad\qquad y = \dfrac{35}{9}$

3. $\quad \dfrac{3}{4y} = \dfrac{16}{3}$ $\qquad$ We must move y from the denominator to the numerator

$\dfrac{y}{1} \times \dfrac{3}{4y} = \dfrac{16}{3} \times \dfrac{y}{1}$ $\qquad$ we have eliminated y from the denominator, by dividing it into itself It has becomes a numerator on the opposite side

$\dfrac{3}{4} = \dfrac{16}{3} y$

We now eliminate the coefficient of y

$\dfrac{3}{16} \times \dfrac{3}{4} = \dfrac{16}{3}y \times \dfrac{3}{16} = y$ $\qquad$ by division and multiplication

$\dfrac{9}{64} = y$ $\quad$ *vice versa* $\quad y = \dfrac{9}{64}$

4. $4x - 7 = -19$ $\quad$ You're aiming for x = something. You know that the Equation Rule says you can do anything you want to one side providing you do the equivalent to the other side. The key words to one side. In this situation you have two obstacles; $4x$ *and* -7. To make 4x become x we would have to multiply both sides by ¼. This would also affect the -7 which is a complication. The -7 means subtract -7 from the left hand side, that's an easy thing to eliminate. Just add +7 to both sides:

$$4x - 7 = -19$$
$$+ 7 = + 7$$

We now have 4x = –12   Eliminating the coefficient of x we have

$$\frac{1}{4}4x \ = \ -12\frac{1}{4}$$

$$x \ = \ -3$$

5. $\frac{2}{3}x \ - \ 12 \ = \ -19$

$$+ \ 12 \ = \ + \ 12 \qquad \text{Equation Concept}$$

$$\frac{2}{3}x \ = \ -7 \qquad \text{Equation Rule}$$

$$\frac{3}{2} \times \frac{2}{3}x \ = \ -\frac{7}{1} \times \frac{3}{2} \qquad \begin{array}{l}\text{Equation Rule and Rule 2 of Three One} \\ \text{Rules}\end{array}$$

$$x \ = \ -\frac{21}{2}$$

6. $\frac{4}{5}x \ + \ \frac{4}{3} \ = \ \frac{3}{5}$

$$-\frac{4}{3} \ = \ -\frac{4}{3}$$

$$\frac{4}{5}x \ = \ \frac{3}{5} - \frac{4}{3} \ = \ \frac{3}{5} \times \frac{3}{3} - \frac{5}{5} \times \frac{4}{3} \ = \ \frac{9-20}{3\times5} \ = \ -\frac{11}{15}$$

$$\frac{5}{4} \times \frac{4}{5}x \ = \ -\frac{11}{15} \times \frac{5}{4} \ = \ -\frac{11}{15} \times \frac{5}{4} \ = \ -\frac{11}{12}$$

$$3 \times 5$$

$$x \ = \ -\frac{11}{12}$$

7.

$$\frac{\frac{3}{4}x - \frac{5}{6}}{\frac{5}{12}} = 10$$

Eliminate the $\frac{5}{12}$ by multiply both sides by $\frac{5}{12}$

$$\frac{\frac{5}{12}}{1} \times \frac{\frac{3}{4}x - \frac{5}{6}}{\frac{5}{12}} = 10 \times \frac{5}{12}$$

Equation Rule, $\frac{\frac{5}{12}}{1} \times \frac{5}{12}$

$$\frac{\frac{3}{4}x - \frac{5}{6}}{1} = \frac{10}{1} \times \frac{5}{12} = \frac{2 \times 5 \times 5}{6 \times 2} = \frac{25}{6}$$

$$\frac{3}{4}x - \frac{5}{6} = \frac{25}{6}$$ There's quite a bit of #3 Rule

$$+\frac{5}{6} = +\frac{5}{6}$$

$$5 \times 6$$

$$\frac{3}{4}x = \frac{25}{6} + \frac{5}{6} = \frac{30}{6} = \frac{5}{1}$$

$$\frac{4}{3} \times \frac{3}{4}x = \frac{5}{1} \times \frac{4}{3} = \frac{20}{3}$$

$$x = \frac{20}{3}$$

8. $\frac{3}{8}x - \frac{7}{2} = \frac{5}{16}x + \frac{15}{4}$ **The major problem, *x* is on both sides** We have to determine which side has the largest groups of x and eliminate the x from the smaller side, Make denominators the same.

$$\frac{2}{2} \times \frac{3}{8}x - \frac{7}{2} = \frac{5}{16}x + \frac{15}{4}$$

We have only multiplied by 1

$$\frac{6}{16}x - \frac{7}{2} = \frac{5}{16}x + \frac{15}{4}$$

element $\frac{6}{16}x$ larger by a value of

$$-\frac{5}{16}x = -\frac{5}{16}x$$

$\frac{1}{16}$ subtract $\frac{5}{16}$ both sides

$$\frac{1}{16}x - \frac{7}{2} = +\frac{15}{4}$$

$$+\frac{7}{2} = +\frac{7}{2}$$

$$\frac{1}{16}x = \frac{15}{4} + \frac{7}{2} = \frac{15}{4} + \frac{7}{2} \times \frac{2}{2} = \frac{15}{4} + \frac{14}{4} = \frac{29}{4}$$

$$4 \times 4$$

$$\frac{16}{1}\frac{1}{16}x = \frac{29}{4} \times \frac{16}{1} = 116$$

$$x = 116$$

Solving many problems, which only require short rote memory procedures, which only work on a specific problem, and develop little understanding, only make you more proficient on these types of problems.

It's only through solving more varieties of challenging problems, using with understanding, mental tools, such as Concepts, Rules, Reflection, visualization, and Pattern recognition which will help prepare you to be a problem-solver with ability and confidence to solve future real life logical problems. Solving the mathematics problems in this book has demonstrated that with understanding you can solve many different types of problems, using ideas and mental tools that you used on previous problems. Concept, Rules, and Patterns impart understanding, they are transferable and can be used in completely different situations. You have only needed to put the ideas together in a slightly different way. The habit that you have been developing of Analysing What type of problem is this? How am going to solve it? Why am I using the Concepts and Rules I'm using. Why am I Thinking, Feeling, Behaving this way; try to make sure your Thinking and Feeling is for the right reason, honesty, trustworthiness, kindness, the virtues. If your behaviour is not for the right reason, why not. Your behaviour should always be a good example for others to follow.

It's the variety of problems that you have previously done, the complexity, the number of steps needed to solve the problem, and the developed understanding, which determine your ability to problem-solve unfamiliar problems. If you have got this far in my book. You have already demonstrated, you've got good problem-tools and the ability to use them. It's up to you to continue practicing, solving logical problems, but always with understanding.

Currently, I feel it's important to share a piece of advice for the wise. Being an excellent logical problem-solver may not fully prepare you for situations that demand subjective methods of thinking.

Life often presents us with challenges that are neither black nor white, but shades of grey, full of complexity and nuance. As a skilled problem solver, you likely trust your ability to make sound decisions. Being an excellent logical problem-solver may not fully prepare you for situations that demand subjective methods of thinking.

When faced with scenarios that require subjective thinking— empathy, emotional understanding, thoughtfulness, or togetherness—relying solely on logic can lead to misguided outcomes. You may believe you're making fact-based decisions when, in reality, emotions—your own or others'—may be influencing your choices more than you realize.

This doesn't only apply to interpersonal situations. It could also involve decisions about purchases, actions, or other personal matters. You're not a machine; you're a human being, inherently prone to the influence of emotions and biases. Every day, we face the same dilemma: choosing between what we should do and what we would like to do.

In such moments, let's hope that virtues like empathy, care, and thoughtfulness guide our decisions, rather than relying solely on rigid, logical reasoning.

CHAPTER 9

Linear Equations and the Cartesian Coordinate System of Graphing.

In Chapter 8 we studied first order equations involving one algebraic variable (one unknown). We are now going to investigate first order equations involving two algebraic variables (two unknowns), usually identified as "x" and "y". They involve situations where the value of one variable the x, causes a change in the other variable the y. They are called linear relationships. They are more engaging and offer more higher-level thinking opportunities. They give more goal orientated problem-solving experience, enabling the acquiring of knowledge, and understanding, using visualization, reflection, and selective attention.

The General Equation for a first order Linear Equation is always of the form:

$$y = ax + b$$

Where a and b are always constant in value; x and y are algebraic variables (the two unknowns); and the value of y is a function of the value of x.

When an equation is written as above, the algebraic variables are given discrete names. The variable on the left (the y) is called the dependent variable. It's called the dependant variable because its value depends on the value of x. Because "a" and "b" are always constants if "x" changes "y" will also have to change to keep both sides of the equation identical in value. It's the variable on the right, the "x", that initiates the change resulting in a new value of y. There is always a relationship between the two variables. They are called an ordered pair and expressed as (x,y). This is why the use of subscripts is always useful. You always know which x and y go together . In trial one you would use $(x1,y1)$. For trial two you would use $(x2,y2)$, the subscript identifies the variable in each trial.

We can see from the relationship, each time x is given a new value, the value of the whole right side of the equation would change. This means the left side (y) must change by an equal amount. This type of relationship occurs frequently in everyday activities.

What is the effect on "y" when "x" changes value? This is something we need to know and something we can investigate. We are looking for a pattern in the relationship of how "y" changes with a change in "x". Could we better increase our understanding of how a change in "x" changes the value of "y" in the General Equation if we change the value of "x" by the same amount each time? If we are changing "x" by the same amount that is a pattern. Would that mean "y" would also change by some constant amount? That would be a pattern. That's something we can investigate.

We must investigate this scientifically. We will perform five trials, each consisting of a new pair of ordered (x,y)pairs that we will use in our calculations. We'll change the "x" value by a constant discrete amount and see if "y" also changes by a constant discrete amount.

If we are going to change the value of "x" by the same amount each time, we have a simple Mental Tool to make things easier. In Mathematics we have a universal symbol for "*change in*", the Greek alphabet letter *delta ("Δ")*. "Δ" means "*change in*".

In our investigation we will use our new delta symbol,
Δx = *the change in x*.

We can also use our previous idea of subscripts:

x1 (*x* sub one) to represent the value of "*x*" in trial 1,
x2 (*x* sub one) to represent the value of "*x*" in trial 2, and so on.

Likewise, we could identify the resultant values of "y" the same way:

When $x = x_1$. then $y = y_1$.
                 $x = x_2$ then $y = y_2$ and so on.

If we are looking for a pattern in values of "*y*", then it makes sense, and necessary to start with a pattern for our values of "*x*". *Using the General Linear Equation,* **y = ax + b**. We will perform five trials, keeping Δx **constant** to determine if there is a pattern in the values of "*y*". See Fig 1. Next page

Fig. 1 Five trials. An investigation of $y = ax + b$ keeping **Δx** constant

We will arbitrarily use: **Δx = 3** $x_1 = 2$ $a = 2$ $b = 7$ and follow the pattern of $y = ax + b$ the General Equation of a Linear Equation

Starting with. $y = ax + b$ (1)
Substituting $x = x_1$ and
$y = y_1$ in equation (1) gives us $y_1 = ax_1 + b$ (2)

Applying our given values of $x_1 = 2$, $a = 2$ and $b = 7$ in Equation (2) and simplifying. we have.

$y_1 = 2 \times 2 + 7$

$y_1 = 11$ (3)

Calculating the value of x_2

$x_2 = x_1 + \Delta x = 2 + 3 = 5$

$y_2 = ax_2 + b$

$y_2 = 2 \times 5 + 7$

Simplifying

$y_2 = 17$ (4)

Calculating the value of x_3

$x_3 = x_2 + \Delta x = 5 + 3 = 8$

$y_3 = ax_3 + b$

$y_3 = 2 \times 8 + 7$

Simplifying

$y_3 = 23$ (5)

Calculating the value of x_4

$x_4 = x_3 + \Delta x = 8 + 3 = 11$

$y_4 = ax_4 + b$

$y_4 = 2 \times 11 + 7$

Simplifying

$y_4 = 29$ (6)

Calculating the value of x_5

$x_5 = x_4 + \Delta x = 11 + 3 = 14$

$y_5 = ax_5 + b$

$y_5 = 2 \times 14 + 7$

Simplifying

$y_5 = 35$ (7)

Notice the pattern of communication in the Mathematics language

Clear, Correct, and Concise

To check for a pattern, we first calculate the change in the dependent variable "y" that has occurred as a result of changing the value of the independent variable "x". We calculate the difference by subtraction. We subtract the initial values of "x" and "y" from the resultant values obtained from the change. Using values from Fig.1 above:

When $x_1 = 2$ We have $y_1 = 11$

When $x_2 = 5$ We have $y_2 = 17$

When $x_3 = 8$ We have $y_3 = 23$

$$\text{When } x_4 = 11 \qquad \text{We have } y_4 = 29$$
$$\text{When } x_5 = 14 \qquad \text{We have } y_5 = 35$$

We calculate the change in Fig. 2

Fig. 2 Change in the Dependent Variable as a Result of a constant change in the Independent Variable from 5 Trials (Fig 1)

Independent variable　　　　　　**Dependent variable**

$x_2 - x_1 = 5 - 2 = 3$ 　　　Gave us　　　$y_2 - y_1 = 17 - 11 = 6$

$x_3 - x_2 = 8 - 5 = 3$ 　　　Gave us　　　$y_3 - y_2 = 23 - 17 = 6$

$x_4 - x_3 = 11 - 8 = 3$ 　　　Gave us　　　$y_4 - y_3 = 29 - 23 = 6$

$x_5 - x_4 = 14 - 11 = 3$ 　　Gave us　　　$y_5 - y_4 = 35 - 29 = 6$

There is a definite pattern, as the independent variable x changes by 3, the Dependent variable y changes by 6.

At first it takes more effort learning and using the language of Mathematics, and to know the meanings of algebraic variables, but through making this effort you are learning how to use and understand, the Concepts, Rules, subscripts, indices, abstract and concrete symbols, mental and material tools, that Mathematics uses to solve logical problems. You are also learning how to analyse, put ideas together in an organized manner, and have the confidence to use all of these skills in your future everyday activities

It is also very important to realize; The number of ideas you can use at any one time, according to the Theories of Vygotsky, is a measure of how able you are to think at a higher level. You can see how many ideas you are already using to solve problems. This should give you more confidence in accepting "it's not your ability that is lacking in being able to think mathematically, but your lack of understanding of how you use Concepts and Rules in problem-solving".

You have been able to work in your native language (perhaps English) where the same symbols, words, and sentences can have several different meanings, yet through great effort you have been successful in learning that language.

The language of Mathematics is much simpler and much easier to learn. In Mathematics you are learning to express your ideas, thoughts, and feelings. Clearly, Correctly, and Concisely. One symbol, one meaning; one word, one meaning; one sentence, one meaning; one paragraph two or more meanings, but always the same two or more. These ideas are not difficult to learn and will make your learning, and ability to use Mathematical thinking in every day problem solving, much easier.

Arranging what you have learned from our 5 trials into a Table Fig. 3 we have on the next page:

Fig. 3 Table of Results of Five Trials

Change in x $(x_2-x_1) = \Delta x$	x_n	$y = ax + b$ $a = 2 \quad b = 7$ $x_1 = 2 \quad \Delta x = 3$	y_n	Change in y $(y_2-y_1) = \Delta y$
		$y_1 = ax_1 + b$		
	$x_1 = 2$	$y_1 = 2 \times 2 + 7 = 11$	$y_1 = 11$	
$(x_2 - x_1) = 3$	$x_2 = 5$	$y_2 = 2 \times 5 + 7 = 17$	$y_2 = 17$	$(y_2-y_1) = 17-11=6$
$(x_3 - x_2) = 3$	$x_3 = 8$	$y_3 = 2 \times 8 + 7 = 23$	$y_3 = 23$	$(y_3-y_2) = 23-17=6$
$(x_4 - x_3) = 3$	$x_4 = 11$	$y_4 = 2 \times 11 + 7 = 29$	$y_4 = 29$	$(y_4-y_3) = 29-23=6$
$(x_5 - x_4) = 3$	$x_5 = 14$	$y_5 = 2 \times 14 + 7 = 35$	$y_5 = 35$	$(y_5-y_4) = 35-29=6$

We reflect on and visualize on our data. We can see having a discrete value for Δx enabled us to see a definite pattern. When we analyse the equation, we see that by increasing x by 3 the value of y increased by 6. Reflecting on the equation, you can see this has to be, as each time you increase x by "3" you have added two threes (the value of $3a$ where "a = 2" to the right side of the equation.

The information in the Table is very useful, but limited as, it only gives information according to the values of x inserted into the equation. We need a bigger more complete picture of what's happening.

More reflection and visualization. Only two elements are changing value in the equation. The algebraic symbols $\underline{x}$ and y. Can you think of any real-life situation where the value of one thing only depends on the value of something else? Yes, your job; how much you earn depends on how many hours you work.

Let's start with an even simpler problem, one working with two values that change but are independent of each other. One you've had to solve when you first started school. In this problem the value of one thing didn't affect the value of the other (one value wasn't dependent on the other), but two values are changing, and the result will be a discrete value (position). Perhaps this similar problem, which we are familiar with, well help us with our Linear Equation problem as both situations have two variables.

How did you find your desk in the classroom? Could you communicate to a friend exactly where your desk is? Could you do this with only two pieces information?

That's an interesting possibility. That's something you've done many times. It's a problem-solving learning situation worth pursuing. We will learn a lot from the investigation, if:

- We use our present knowledge and understanding, and
- Our What, How, and Why Mathematical thinking.

Starting with Reflection, Prior Knowledge, and Visualization.

WHAT do you know about the classroom?

- Could a drawing with all of the relevant information help?
- We're using Maths so our information must be exact.
- That means we are working with discrete pieces of information.
- What key elements can you say about the classroom?

You reflect and visualize. The classroom floor is a rectangle. Opposite sides are equal and parallel, the sides make four equal corners which are 90 degrees (We have another mental tool to indicate 90 degrees, the perpendicular symbol "⊥". We say the sides are ⊥ (Perpendicular) to each other. There are two doors, one in the front corner and another at the opposite corner at the back of the classroom.

The desks are arranged in straight lines, which make columns, and rows. The rows are parallel to the front and back of the classroom, the columns are parallel to the sides of the classroom, therefore the rows are ⊥ to the columns. There is a space at the front of the classroom for the teacher's desk, and there are pathways between the rows as well as between the columns. A diagram of your classroom showing your desk is shown in **Fig. 4** on the next page.

Fig. 4 Classroom

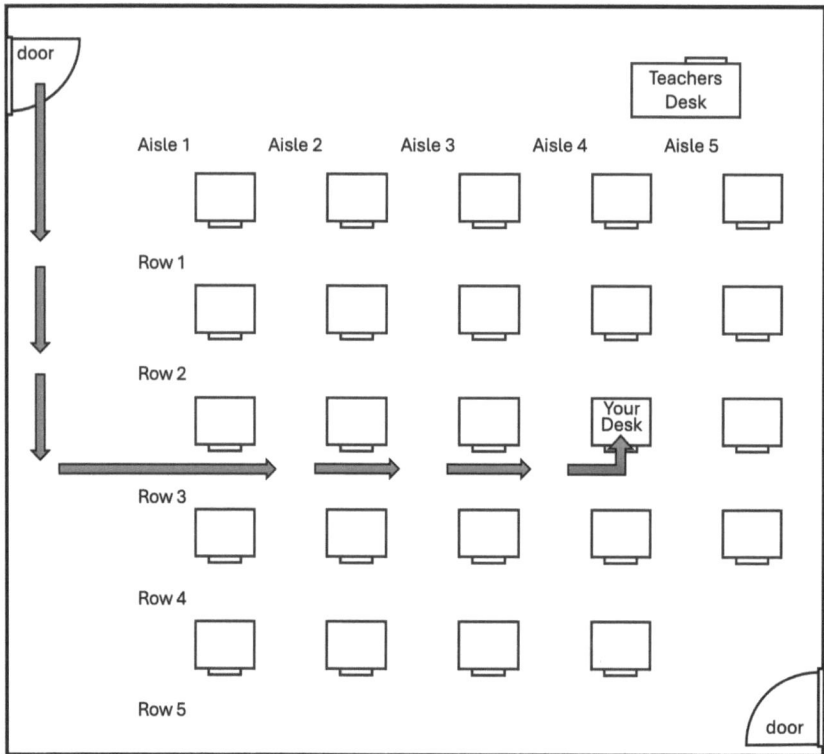

How would you communicate to someone where your desk is? Would you have them take the shortest path, a straight line from the door to your desk, expecting them to climb over the desks. Of course not. Could they reach your desk following a single straight line. No, only if your desk is the closest one to the door The shortest route would require two straights $\perp$ lines.

Given your desk is in Row 3, Aisle 4 you'd walk down Aisle 1, until you reached Row 3, make a 90-degree left turn, continue your walk along Row3, until you reached Aisle 4, where your desk was located. You have needed to walk along two perpendicular lines to reach your desk.

We've designed a system which works well for the above classroom, however not all classrooms are the same. Could we generalize our system so it would work for any classroom, and also lead to a system that would work in the general equation, which uses two unknowns, with the value of the first, dependent on the value of the second? This will be our first big challenge. Let's see how we do.

First, not all classrooms enter by the front door, the entrance could be by the back door. We'll have to modify our system so that it can work, for either front, or back. We'll have to have universal ideas. We're either going up or down or right or left. We also need to realize that up or down, or right or left means we need a transition point, when you are neither up or down or left or right. You are at a zero position of movement.

If we are to Generalize our ideas we must have simple Universal symbols, Concepts, and Rules. We'll define the Following:

- Moving left or right is moving in the x Direction.
- Moving to the right is positive and to the left is negative.
- The position "$x = 0$" defines neither left or right.
- Moving up or down is moving in the y Direction
- Moving up is positive and moving down is negative.
- The position $y = 0$ defines neither up nor down.

To clarify and reinforce the above we'll define moving to:

- the right ⇒ or up ⬆, as positive ➕ movement, and

- moving to the left ⬅ or down ⬇ as negative ➖ movement.

See Fig. 5A below for an example of a set of x Lines to measure the right or left position. Likewise, we will need a reference Line where by definition, you are neither left nor right. This is the **x = 0.** By definition the **x = 0** line is also called the **y-axis**

x-Lines are Vertical Lines   y-axis   **Fig. 5A**

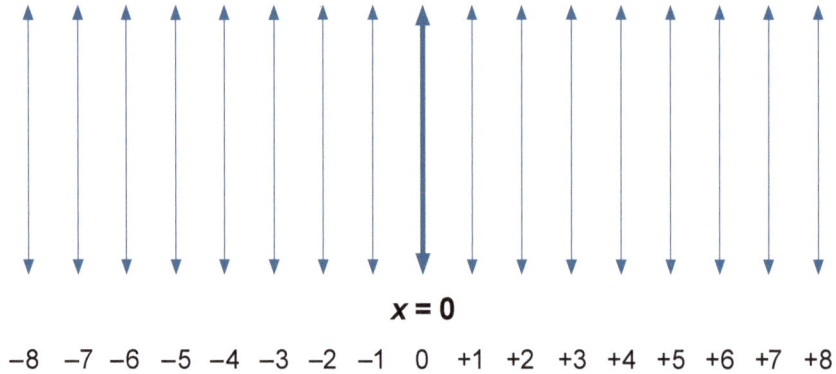

$$x = 0$$

−8 −7 −6 −5 −4 −3 −2 −1   0 +1 +2 +3 +4 +5 +6 +7 +8

See Fig. 5B below for an example of a set of y Lines to measure the up or down position. Likewise, we will need a reference Line where by definition, you are neither up nor down. This is the **y = 0** Line. The the **y = 0** line is also called the **x-axis.**

y-Lines are Horizontal Lines   **Fig. 5B**

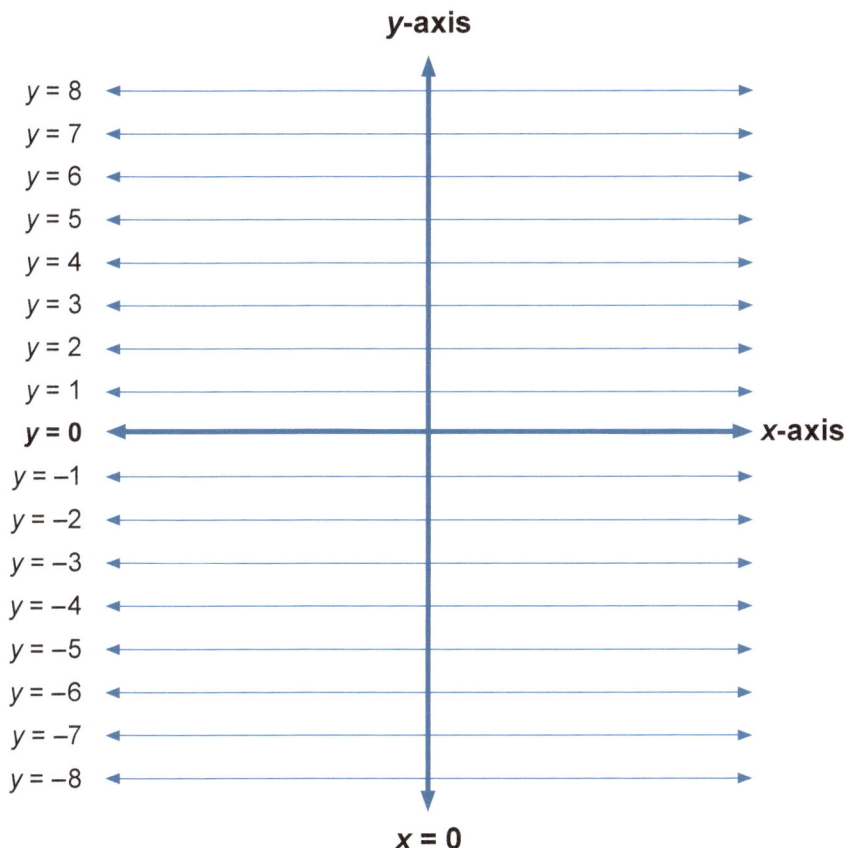

y-axis

y = 8
y = 7
y = 6
y = 5
y = 4
y = 3
y = 2
y = 1
y = 0   **x-axis**
y = −1
y = −2
y = −3
y = −4
y = −5
y = −6
y = −7
y = −8

x = 0

The x or y Line only measures the distance you are from the $x = 0$ or y = 0 Line. It's only when Lines intersect that a definite position is created.

We are trying to create a system to fit the General Equation $y = ax + b$. Our "x" and "y" values can be either positive or negative, so it makes sense to combine both sets of straight-line systems. This is done below. See Fig 6

Fig. 6   **Cartesian Graph**

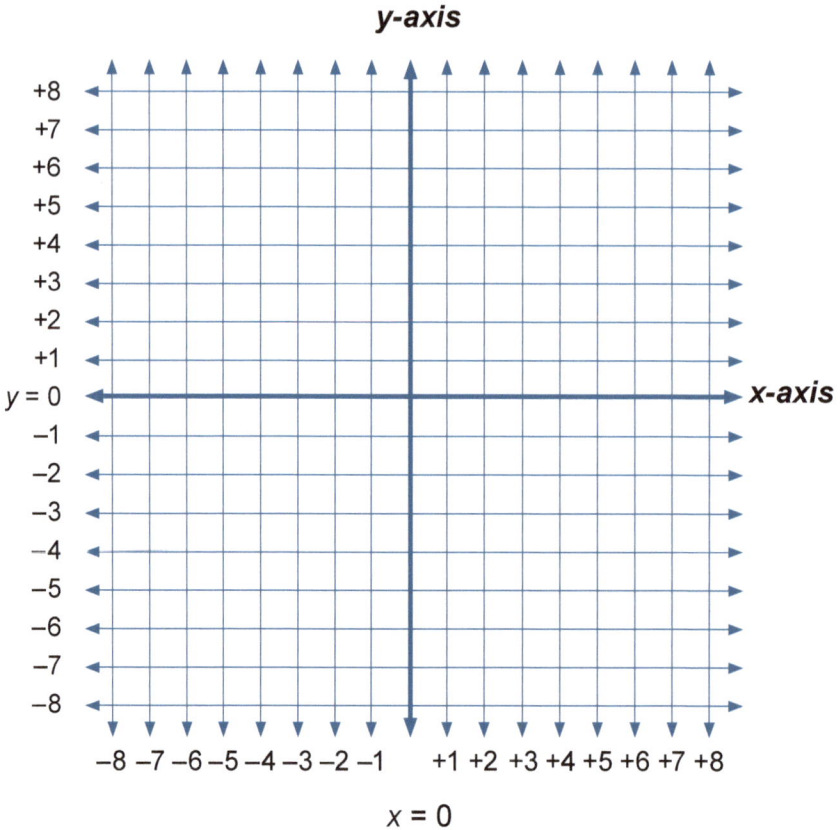

By overlapping the two different sets of x and y Lines they have become ⊥ to each other. In doing so we have created a very powerful new Mental Tool in Mathematics. It's called a Cartesian Graph named after René Descartes (1596–1650).

This powerful tool enables you to plot, on the graph, the two values you use, or obtain in a two-solution problem situation. It gives you a visual view. A view that might expose a possible pattern of behaviour between the two values.

In the classroom you were always constrained by the number or rows and aisles. You were given both the row and aisle enabling you to find your desk. Each aisle line could take you to several different number of rows.

In the General equation, for each value of x there is only one value of y. These two values together are called an ordered pair and presented as an (x,y) pair. The x value always presented before the y value. The value y which you obtain is always dependent on the value of x you used. If the values of the ordered pairs used were plotted in a graph, and the points joined together, in ascending or descending order, any pattern produced on the graph could be observed. This gives you another means of gaining more understanding and information towards how algebraic variables interact when in an equation.

Following are the properties we now know about the "x" Lines and "y" Lines when they are in a Cartesian Graph:

- The "x" and "y" Lines are always $\perp$ to each other.

- The y-axis and the $x = 0$ Line occupy the same Line.

- The x-axis and the $y = 0$ Line occupy the same Line.

- The "x" value measures the distance in "x" units away from the y- axis ($x = 0$ Line). The y–axis is used because it is the reference line for moving the perpendicularly distance.

- The "y" value measures the distance in "y" units away from the x– axis ($y = 0$ Line). The x–axis is used because it is the reference line for moving the perpendicularly distance.

- The spaces between the x-Lines and y-Lines measure the fractional values of the x and y variables.

- When the line that has been drawn through the solution points crosses the x-axis, that point is called the x-intercept (x_{int}). To be on the x-axis. the y value must be equal to zero.

- When the line that has been drawn through the solution points crosses the y-axis that point is called the y-intercept (y_{int}). To be on the y-axis the x value must be equal to zero

- When you only have the value of the "x" variable your position is somewhere on a line parallel to, and x units away, from the y-axis

- When you only have the value of the "y" variable your position is somewhere on a line parallel to, and y units away, from the x-axis.

To reinforce an important point, the values of the x and y variables are always given as an ordered pair with the x value given first. They are presented as an (x,y) pair

When you are given an x value you are really talking about an x Line, likewise when you are given a y value you have a y Line. When you have an ordered pair (x,y) you are talking about the intersection of the x and y Lines. When lines intersect, they create a point. When working with a point that point is a position. Each ordered pair we create gives us a new position.

If we draw a line through adjacent points, a useful pattern might be created, depending on the closeness of how many plots we have. If there is a pattern, it would show a clear picture of the relationship between the equation's dependent, and independent, variables. A clear picture gives us another mental tool for increasing our understanding of Linear Equations.

We can now use this new tool, the Cartesian Graph. We will plot the ordered pairs, solutions we obtained from our first Linear General Equation, , y = 2x + 7. Recorded in **Fig 3 page 191.** They will give us positions on the Graph.

For ease of access, **Fig. 3** is repeated below

Fig. 3 Table of Results of Five Trials

Change in x $(x_2-x_1) = \Delta x$	x_n	$y = ax + b$ $a = 2 \quad b = 7$ $x_1 = 2 \quad \Delta x = 3$	y_n	Change in y $(y_2-y_1) = \Delta y$
		$y_1 = ax_1 + b$		
	$x_1 = 2$	$y_1 = 2 \times 2 + 7 = 11$	$y_1 = 11$	
$(x_2 - x_1) = 3$	$x_2 = 5$	$y_2 = 2 \times 5 + 7 = 17$	$y_2 = 17$	$(y_2-y_1) = 17{-}11{=}6$
$(x_3 - x_2) = 3$	$x_3 = 8$	$y_3 = 2 \times 8 + 7 = 23$	$y_3 = 23$	$(y_3-y_2) = 23{-}17{=}6$
$(x_4 - x_3) = 3$	$x_4 = 11$	$y_4 = 2 \times 11 + 7 = 29$	$y_4 = 29$	$(y_4-y_3) = 29{-}23{=}6$
$(x_5 - x_4) = 3$	$x_5 = 14$	$y_5 = 2 \times 14 + 7 = 35$	$y_5 = 35$	$(y_5-y_4) = 35{-}29{=}6$

The ordered pairs are plotted in the Cartesian Graph **Fig. 6** on the next page. .

Fig. 7 Linear Equation y = 2x +7

Plot of 5 ordered pairs gained from the 5 trials, plus two plots by substituting $x = 0$, and then $y = 0$ to show the x and y intercepts. The dots were in alignment allowing a straight line to be drawn through all of the ordered pairs.

y - axis

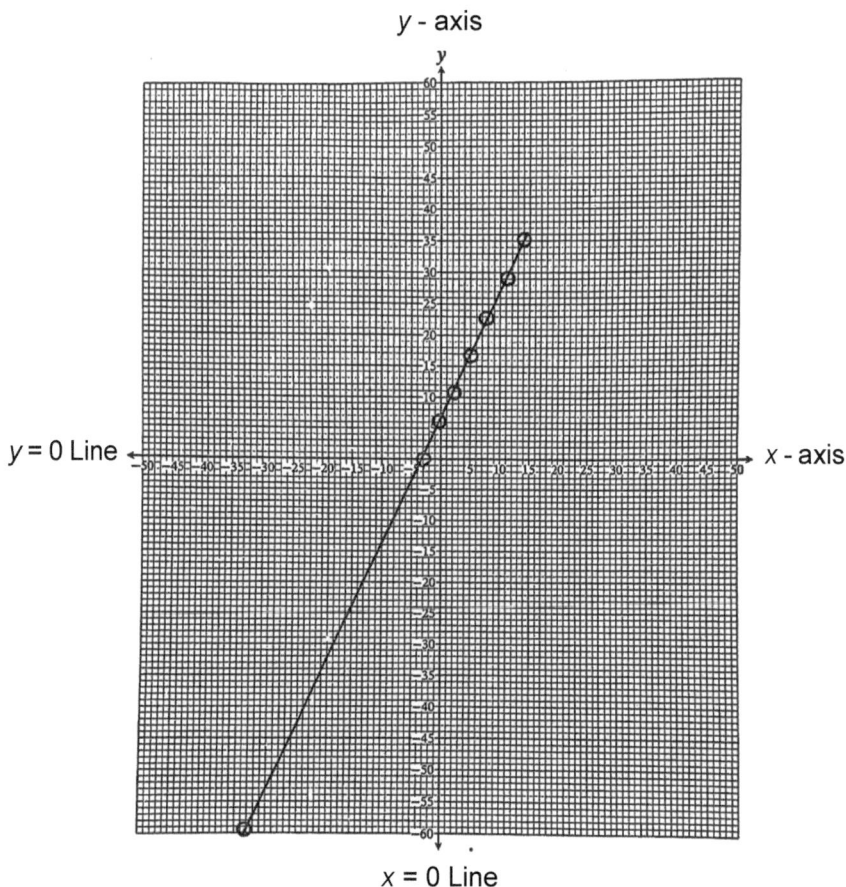

y = 0 Line

x - axis

x = 0 Line

An observation of the graph shows that a straight line can be drawn through the first seven points. If we continue the straight line to the bottom line of the graph it crosses the y = −60 Line. That position is a significant separation from the point on the y = 0 Line. If we calculate the x position for the y = −60 Line, and find that point lies on the line, it will reinforce our belief that the solution set for y = 2x +7 is a straight line.

We will investigate to see if y = 60 is part of the ordered pair which is on the line.

Starting with our equation	$y = 2x + 7$	(1)
Substituting y = −60		
into equation (1) we get.	$-60 = 2x + 7$	(2)

$$-60 = 2x + 7$$

Adding −7 to both sides of

$$-7 = -7$$

Equation (2) we obtain

$$-67 = 2x \qquad (3)$$

Dividing both sides of equation
(3) by 2 and reversing both sides
gives us equation $\qquad\qquad x = -33.5 \qquad$ (4)

Our ordered pair becomes $\qquad$ (−33.5, −60) $\qquad$ (5)

When we plot (−33.5, −60) into the graph on page 192 it lies on the line. We feel confident that the solution set is a straight line. But "Feel confident" is not a fact! In Mathematics we are always working with facts We will have to do more reflection and visualization.

A major piece of information about a straight line is its **Slope.** The slope is the angle a straight line makes with the x-axis. It is defined by the equation:

$$slope = \frac{\Delta y}{\Delta x} = \frac{y_2 - y_1}{x_2 - x_1}$$

Let us start with a graph **Fig.7** made from an unknown linear equation, as given on the next page.

Fig. 8 Determining the slope and linear equation from a Graph

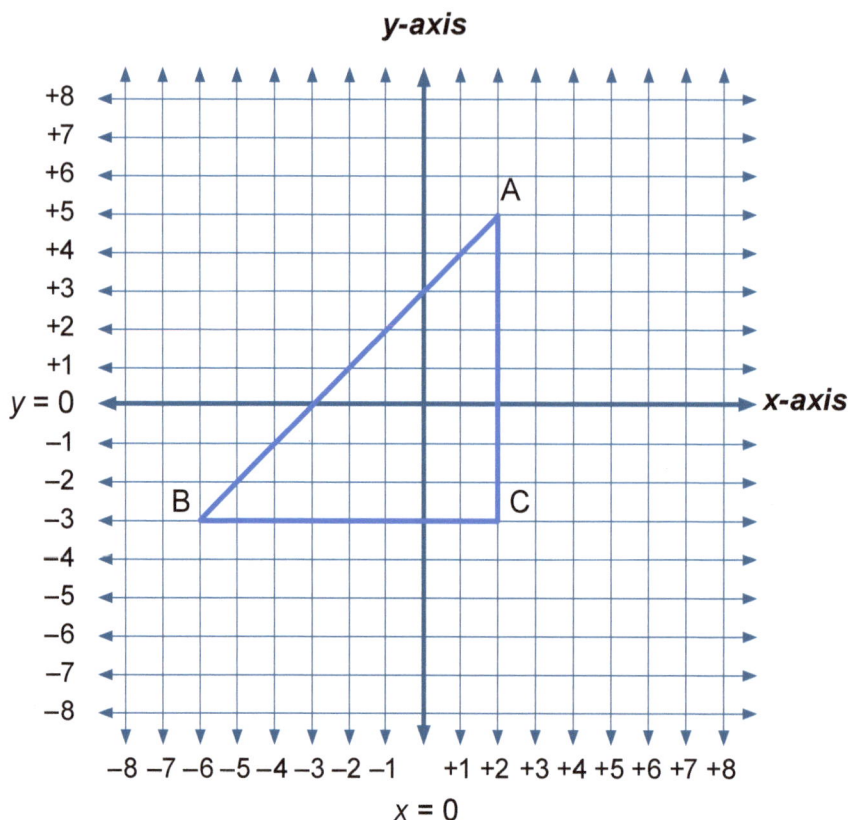

$$slope = \frac{AC}{BC} = \frac{y_2 - y_1}{x_2 - x_1}$$

Vertex A is (2,5), C is (2,–3) and B is (–6, –3)

It is easier to visualize $(y_2 - y_1)$ by realizing A, B and C are the vertices of the triangle.

Side AC = $(y_2 - y_1)$ and side BC = $(x_2 - x_1)$.

We can express the actual lengths of the sides AC and BC of the triangle as:

AC = 5 – (–3) = 8 and

BC = 2 – (–6) = 8

Substituting the Values for AC and BC we obtain

Slope $= \dfrac{\text{AC}}{\text{BC}} = \dfrac{y_2 - y_1}{x_2 - x_1} = \dfrac{8}{8} = \mathbf{1}$

We are now ready to attack the problem of constructing the linear equation which describes the line.

We start with what we now know:

- **y = ax + b,** the General Equation,
- the vertex points **A** (2,6) and **B** (–6, –3),
- the vertex points are on the line and are ordered pairs.
- Slope = 1

Our first issue is the General Equation, It has two constants **a** and **b** and the two algebraic variables x and y, that's four unknowns that we will have to work with. That's too many.

We'll have to break the problem down into manageable parts.

We can use the two sets of ordered pairs for **A** and **B**. They are part of the line and are contained in the solution set for the equation we are looking for.

They are two different points (two different ordered pairs). That means using the General Equation we can generate two different useful equations.

We've divided the problem into two smaller parts. We will generate the two equations and see how that can help

Starting with our equation $\qquad\qquad y = ax + b \qquad\qquad$ (1)

Substituting the ordered pair
values for **A**, (2,5) in equation (1)
gives us our first equation $\qquad\qquad$ **5 = 2a + b** $\qquad\qquad$ (2)

Substituting the ordered pair
values for **B**, (−6, −3) in equation
(1) gives us our second equation $\qquad$ **−3 = −6a + b** $\qquad$ (3)

This has given us two equations each containing the two constants, **a** and **b** that are contained in the equation we're looking for

We are in a position of strength. If we can eliminate the constants one by one by manipulating the two equations in some way, we will end up finding the values of **a** and **b** that we're looking for.

If we multiply either equation by "−1", we can then add the resultant equations together and that would eliminate **b** leaving us with a single simple equation with **a** that we know we can solve. We learned to do this when we worked with simple equations.

If we multiply both sides of
Equation (3) by −1 we obtain $\qquad\qquad$ 3 = 6a − b $\qquad\qquad$ (4)

Adding equations (2) and (4) $\qquad\qquad$ 5 = 2a + b $\qquad\qquad$ (2)
together eliminates constant **b** $\qquad\qquad$ 3 = 6a − b $\qquad\qquad$ (4)

giving us a solvable equation (5) $\qquad\qquad$ 8 = 8a $\qquad\qquad$ (5)

Dividing both sides of Equation
(5) by 8 and rotating the sides
we obtain the value of our first $\qquad\qquad$ a = 1 $\qquad\qquad$ (6)
constant

Substituting our value for a in either equation

Substituting in equation (3). $-3 = -6 + b$

Simplifying we obtain $b = 3$

We can now insert our obtained values of a =1 and b = 3 in the equation of the line y = ax + b. This gives us

$$y = x + 3$$

It's always a good idea to check our result. We can easily do this by going back to the **Fig. 7** graph. We extend our given line to the y = 8 Line. We notice that it intersects the x = 5 Line. This gives us the ordered pair 5,8) which is position D. We will insert this ordered pair into our equation to see if it is a solution to the equation.

$$y = x + 3$$
$$8 = 5 + 3$$

Subtracting 5 from each side we obtain

3 = 3. Yes (5,8) is a solution to our equation.

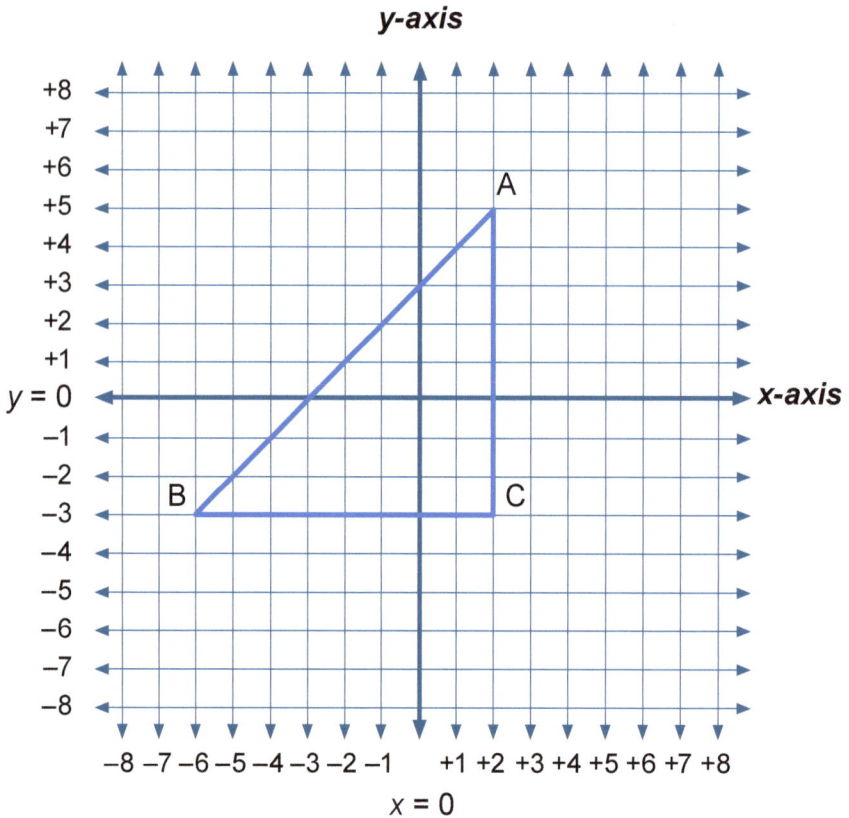

We are always looking for patterns. Let's compare the equation of the line on the graph with the General Equation of a Linear equation to see is there is a pattern.:

Our derived equation $y = x + 3$

The General Equation for a Linear $y = ax + b$

We notice "a" in our derived equation is equal to "1" which is equal to the slope of the line. Is it possible that the constant "a" is always equal to the slope in a Linear Equation. If that's so, it's important information. It's worth investigating.

To prove this. we must demonstrate that in the General Equation "a" is always equal to the slope. If this is true, then this will be true for all linear equations.

Given $y = ax + b$ and that slope is defined as:

$$slope = \frac{\Delta y}{\Delta x} = \frac{y_2 - y_1}{x_2 - x_1}$$

Given $\qquad\qquad\qquad\qquad\qquad y = ax + b \qquad\qquad (1)$

Substituting the order pair (x_1, y_1)

Into equation (1) we obtain $\qquad\qquad\qquad y_1 = ax_1 + b \qquad\qquad (2)$

Substituting the order pair (x_2, y_2)
into equation (1) we obtain $\qquad\qquad\qquad y_2 = ax_2 + b \qquad\qquad (3)$

Subtracting equation (2
From equation (3) we
obtain equation (4) $\qquad\qquad y_2 - y_1 = (ax_2 + b) - (ax_1 + b) \quad (4)$

Simplifying equation (4)
we obtain equation (5) $\qquad\qquad\qquad y_2 - y_1 = ax_2 - ax_1 \qquad (5)$

Collecting like terms in
Equation (5) $(x_2a - x_1a)$
we obtain equation (6) $\qquad\qquad\qquad y_2 - y_1 = a(x_2 - x_1) \qquad (6)$

Dividing both sides of
Equation (6) by $(x_2 - x_1)$
we obtain equation (7) $\qquad\qquad \dfrac{y_2 - y_1}{(x_2 - x_1)} = \dfrac{a(x_2 - x_1)}{x_2 - x_1} \qquad (7)$

Simplifying equation (7) and
reversing both sides
we obtain equation (8). $\qquad a = \dfrac{y_2 - y_1}{(x_2 - x_1)} = \dfrac{\Delta y}{\Delta x} = slope \ (8)$

We found the constant a is always equal to the slope.

If constant a = slope in the General Equation for a Linear Equation, then a is equal to the slope for all Linear Equations.

Let's look at the General Equation for Linear Equations again. Perhaps we can see more patterns. Starting with $y = ax + b$. Using our knowledge of the Cartesian Graph we realize that the x and y values are really lines and the intersection of those lines create a point, which we call an ordered pair. When $x = 0$ you are somewhere on the y-axis. Likewise, when $y = 0$ you are on the x-axis. That means when $x = 0$ you are on the y-axis at the point where the line created by the ordered pair crosses the y-axis. That point is defined as the y_{int} (y intercept). Likewise, when the line crosses the x-axis that point is called the x_{int}. (x intercept). If we apply that knowledge to $y = ax + b$ we have:

$y = ax + b$

If we let $y = y_{int}$ then $x = 0$ Substituting these values in

$$y = ax + b$$

we get $\qquad y_{int} = a \times 0 + b$

Simplifying $\quad y_{int} = b$

Another discovery, just by putting our ideas together in a different way. We have proved; If b = **yint** for the General Equation then:

b = y_{int} for any Linear Equation

You now know quite a lot about Cartesian Graphs but need more practical experience and reinforcement of your understanding. The difficulty arises when you are confronted with two linear equations.

No problem. The solution process involves manipulating the equations so that you end up with only one equation. An equation containing only one unknown, an algebraic variable, or a constant. Once you are in that situation of only one unknown it has become a simple equation. You can always solve a simple equation. That's an equation you've had experience working with.

Summarizing what we have already learned.

- Problem-solving involves using tools
- When the problem is intellectual it is solved with the use of intellectual tools
- The main Intellectual tools include Concepts, Rules, Reflection, and Visualization (Using Why).

Practical experience and reinforcement of your understanding.

In Geometry we know that an infinite number of straight lines can be drawn through one point (ordered pair), but only one straight line can be drawn through two points (Concept in Geometry). Thus, we can define a straight line with two ordered pairs. That means we are looking for the line created from each ordered pair that has the same slope (goes in the same direction) and the same y-intercept (b constant). For this to happen the two lines must superimpose and be identical. The above reasoning means we can set the two lines created by the ordered pair identical to each other because their b constants (y_{int} values) are identical. (2)

Example 1:

Determine the equation of the line which goes through points (7, −3) and (−2,12)

What do we know about the Line?

1. The line goes through two points.
2. The two points define a single line.
3. From the infinite lines (equations) which can be drawn through each point find the one equation which goes through both points.
4. The two lines must have the same slope (a) and the same (b) to be one and same.
5. All linear lines can be expressed by the General Equation, so that's the best way to start our solution process.

We are aiming to have two simple equations. One in **a** and another in **b** which we can use to determine the value of a (slope) and b (the y_{int}).

The equation for any straight line is the General Equation ($y = ax + b$)

Using and inserting the two ordered pairs in $y = ax + b$ we obtain.

$$-3 = 7a + b \qquad (1)$$
$$12 = -2a + b \qquad (2)$$

You can see that there are several ways that we can go to derive our single equation. I have chosen one which gives you more practice manipulating equations.

You can see that both equations contain only one b
If we multiply either equation by -1 then add the two equations together the resultant equation would only have the constant a in it.

Multiplying equation (1) by -1 you get. $\quad 3 = -7a - b \quad$ (3)

Adding equations (2) and (3) together $\quad 12 = -2a + b \quad$ (2)
$$3 = -7a - b \quad (3)$$

you get the simple equation $\quad 15 = -9a \quad$ (4)

dividing both sides of equation (4) by −9,

simplifying and reversing we obtain $\quad a = -\dfrac{5}{3} \quad$ (5)

We now put our effort into finding the value of constant b

Starting with $y = ax + b$ $\qquad y = ax + b \quad$ (1)

Substituting (7,-3) into (1) $\qquad -3 = 7a + b \quad$ (2)
Substituting (-2,12) into (1) $\qquad 12 = -2a + b \quad$ (3)

Multiply both sides of (2) by 2 $\qquad -6 = 14a + 2b \quad$ (6)
Multiply both sides of (3) by 7 $\qquad 84 = -14a + 7b \quad$ (7)

Adding (6) to (7) we obtain $\qquad 78 = 9b \quad$ (8)

Dividing both sides of (8) by 9 then,

Reverse both sides and simplify to obtain $\quad b = \dfrac{26}{3} \quad$ (9)

Inserting our values for a and b into the General Equation,

we obtain the equation of our line $\quad y = \dfrac{-5}{3}x + \dfrac{26}{3} \quad$ (10)

The good news, we can check our Answer.

Substituting (7,–3) into equation (10) $y = \dfrac{-5}{3}x + \dfrac{26}{3}$

We obtain $-3 = -\dfrac{5}{3} \times 7 + \dfrac{26}{3} = \dfrac{-35}{3} + \dfrac{26}{3} = \dfrac{-9}{3}$

Yes (7,–3) is a solution $-3 = \dfrac{-9}{3} = -3$

We already know that the slope of the line is always equal to the constant "a", and that "b" is always equal to y_{int}.
Let's see what else we can discover.

Example 3 In the General Equation for a linear equation we discovered that a is always the value of the slope, and that b was always equal to the y intercept. That's one of the great strengths of Mathematics discovery. Having a media which enables you to get practice putting abstract ideas together in an organized way to come up with a new idea. Also, with more practice and success you gain understanding and confidence to apply these skills in any problem-solving situation.

We know from Geometry that lines which are parallel go in the same direction (Concept in Geometry). That means they have the same slope. If they have the same slope, then the solution sets (lines) of linear equations will have the same constant "a" . Which means all Linear equation which have the same constant "a" are parallel.

Example 2:

Given the two following linear equations

$$y = -\dfrac{2}{13}x - \dfrac{13}{7} \qquad y = -\dfrac{2}{13} + \dfrac{2}{3}$$

Both equations are an identical in their value of "a" That means they are parallel to each other

Given another two linear equations:

$$y = -\frac{2}{5}x - 7 \qquad y = -\frac{2}{5} + 5$$

Their "coefficients x" look the same but they're different . The "a" of one is negative but the "a" of the other is positive. They **are not parallel**. If the slope only differs in sign, how does that affect the situation? Let's investigate. We'll use a graph of the two lines, reflect and visualize is there is any pattern regarding the slope and y_{int} in this situation. It would be best to look for a pattern from the graphs that the two lines make. See below the graph of the two lines.

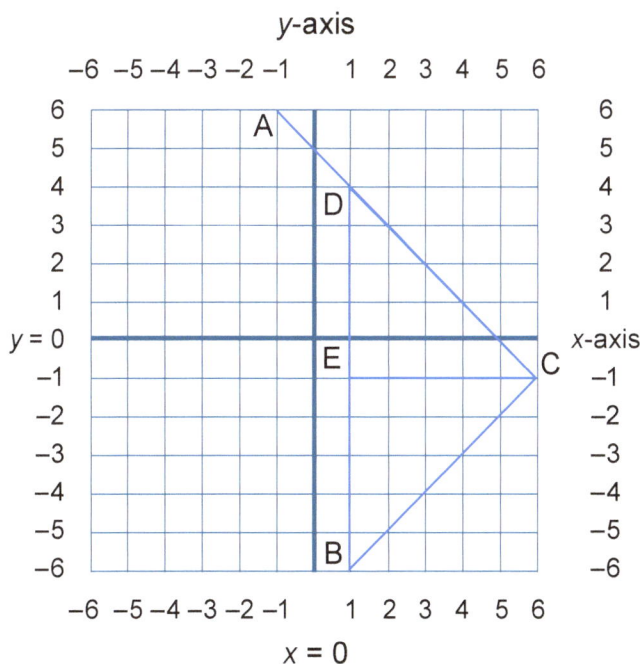

The two lines intersect at point C ((6,-1)

Line AC goes through the ordered pair A (-1,6) and C (6,-1)

Line BC goes through the ordered pair B (1,-6) and C (6,-1)

The slope of AC can be obtained from the Rt angle triangle DEC

The slope of BC can be obtained from the Rt angle triangle BEC

Slope of AC $= \dfrac{ED}{CE} = \dfrac{5}{-5} = -1$

Slope of BC $= \dfrac{EB}{CE} = \dfrac{-5}{-5} = 1$

As can be seen in the graph the slope of either line can be obtained from constructing right angle triangles using the intersecting lines as the hypotonus of the triangles, The lengths of the sides of the right-angle triangle are measured from the intersection point. Because the triangles that are constructed from lines that differ in slope only by one being positive and the other negative the triangles formed will always create right-angle triangles which are similar. The slopes of the two intersecting lines will always have the same value but differ in sign (+ or -).

We have discovered that the x Line and y Line from the ordered pair, (the intersection point), are the angle bisectors of the angles formed by the intersecting lines.

Before we work on the last problem of the chapter, it would be a good idea to have a pause and go over some of the important ideas that we have previously presented,

What are some of the most important things we've learned? Answers supplied at the end of the Revision.

1. A linear equation is a straight line. It can be expressed mathematically as _____ 1)?

2. We use a _____ 2) graph to display its properties.

3. A _____ 3) graph has two _____ 4) straight lines which make an angle of 90 degrees. We say the lines are _____ 5),

4. The two lines are called the _____ 6), and the _____ 7) they measure the backward and forward directions, the _____ 8) direction and the up and down, direction. the _____ 9) direction.

5. Each position on a line is measured by an x value and a y value. These two values together are called an 10) _____ pair.

6. A straight line has three main properties. A _____ 11) the _____ 12) symbol in the General Equation.

7. The _____ 13) value is equal to the "**b**", symbol in the General Equation and an _____ 14) equal to "**-b/a**" symbols in the General Equation.

$$3x - 2 = \frac{2}{3}x + 5$$

8. We also discovered when the line crosses the x-axis, we call the ordered pair of this position (x_{int} , 0). This should be obvious because the x-axis is the. _____ 15) line and x_{int}. **is a unique position**. Likewise, when the line crosses the y-axis, we call the ordered pair of this position (0, y_{int},) This is obvious for the same reason. The y-axis is the _____ 16) line, and y_{int}. is a unique position.

9. When a new or unfamiliar idea is checked and verified by inserting the information into the General Equation, the results are applicable to _____ 17)

Answers

1) y = ax + b   2) **Cartesian**   3) **Cartesian**   4) **intersecting**
5) **perpendicular**   6) *x*-axis   7) **y-axis.**   8) horizontal
9) **vertical**   10) **ordered**   11) **Slope**   12) "**a**"   13) y_{int}.
14) x_{int}.   15) **y = zero**   16) *x* = zero   17) **all Linear Equations**

Example 3

Given the following two lines:

$$y = \frac{2}{3}x + 5 \tag{1}$$

and $\qquad\qquad y = 3x - 2 \tag{2}$

Find their intersection point.

Before we charge right into the problem, we should ask ourselves what do we know about the intersection point?

1. It's an ordered pair that both lines share at the intersection point.
2. At the intersection point you are on both lines at the same time.
3. We can insert the same ordered pair as a solution for each line.
4. We can make two separate equations at the intersect point which have the same value x and y but different constants "**a**" and "**b**".

That means we can set (1) and (2) equal to each over, giving us (3). By doing this we automatically eliminate the y from the equations and are left with (4) a relationship that can easily be solved for *x*.

$$3 \times (3x - 2) = 3 \times \left(\frac{2}{3}x + 5\right)$$

$$9x - 6 = 2x + 5 \tag{3}$$

Simplifying we have: $\qquad 7x = 21 \tag{5}$

Dividing both sides of (5) by 7
We obtain: $\qquad\qquad x = 3 \tag{6}$

We can easily find the value of y by substituting the value of x in (6) in either of our two equations (1) or (2) as at the point of intersecting both equations share the same x and y

Choosing the easiest (2) $\qquad y = 3x - 2 = 9 - 2 \tag{2}$
We obtain

$$y = 7$$

At first it takes more effort learning and using the language of Mathematics, and to know the meanings of algebraic variables, but through making this effort you are learning how to **use** and **understand**:

- the Concepts.
- Rules.
- subscripts.
- indices.
- abstract and concrete symbols.
- mental and material tools.
- What, How, and Why, thinking.

In problem-solving Mathematics uses all of the above to solve logical problems. It's **only through understanding** that you are enabled, and have the self-confidence, to transfer what you have learned for solving problems in entirely different situations. The number of ideas you can use at any one time during a single problem-solving activity, according to the Theories of Vygotsky, are a measure of your ability to think at a higher level.

You can see how many ideas you are already using to solve problems. This should give you more confidence in accepting **"it's not your ability that's lacking in being able to think mathematically, but your lack of understanding".**

You have been able to work in your native language (perhaps English) where symbols, words, and sentences can have a number of different meanings, yet through great effort you have been successful in learning the language. The language of Mathematics is much simpler and much easier to learn. In Mathematics you are learning to express your ideas, thoughts, and feelings. Clearly, Correctly, and Concisely. These ideas are not difficult to learn and will make your learning, and ability to use Mathematical thinking in every day problem solving, much easier.

Composing an Index for this book was very difficult. The study of Mathematics is a study in understanding. One's Understanding is not a discrete place in time but continually grows as new ideas and discoveries are interwoven with the original idea. As you read through this book you looked for patterns which were later remoulded to fit new observations. The result: the same idea was used many times in different situations. It became very difficult to document a particular page for a single idea as it was woven and used in many different ways and places. That means an accurate index of where each idea originated, and how it was used in different situations, would fill many pages. As well it's important that the whole volume is thoroughly studied, to gain the understanding and competency in problem-solving that can be gained from this book. I hope the above discussion explains the brevity of the Index.

Index

Page

References

The writing of this book started in the 1980's and finished in February 2025. During this time my ideas were trialled in the classroom. I have shared my ideas with many people to test the ideas out. It's impossible to recall the books and Journals that I have read during this time. The only Theories that stand out and are mentioned in this book are the Theories of Vygotsky which I discovered in 2012. I would suggest Vol 1 and 2 of The Collected works of L. S. Vygotsky. A more detailed easier to read book which I recommend would be The neo-Vygotskian Approach to Child Development 2005 by Yuriy V. Karpov which only recently came to my notice. It is very difficult trying to find published works which are in tune with your own when you are challenging the deeply entrenched ideas that are already accepted. The very early examples of published papers I have presented are titled "Advance and Justify", and "A metacognitive, Conceptual, Rule-based, Problem-solving, Approach to Mathematics and Problem-solving. Some of my presented papers are still on the Internet.

Brophy, J. (1986). Teaching and Learning Mathematics: Where Research Should be Going. Journal for Research in Mathematics Education, Vol. 17(5), 323-346.

Bloom, B.S. (1976). Human Characteristics and School Learning. N.Y.: McGraw -Hill.

Department of Education (1987). Mathematics Achievement in New Zealand Secondary Schools: A Report on the Conduct in New Zealand of the Second International Mathematics Study with the International Association for the Evaluation of Educational Achievement. Wellington, N.Z.: Government Printer.

De Bono, E, (1976). Teaching Thinking. London: Billing and Sons Ltd.

Dewey, J. (1933). How We Think. Boston, Mass.: D.C. Heath and Co. Duncan, E.R.; Capps, L.R.; Dolciani, M.P.; Quast, W.G.; & Zweng, M.J. (1980). Modern School Mathematics Structure and Use. Boston, Mass: Houghton Mifflin Co.

Firestone, A.H. (1988). A Game-Based Approach to Teaching Problem Solving and Thinking Strategies to Form 1 Intermediate School Pupils Solving Fractions and Algebraic Equations. Paper presented at 10th N.Z.A.R.E. Conference, Massey University, Palmerston North (Dec. 1988).

Firestone, A.H. (1989a). A Meaningful, Rule-based Approach, to Teaching Mathematics. Paper presented at 11th N.Z.A.R.E. Conference, Central Institute of Technology, Heretaunga (Dec. 1989).

Firestone, A.H. (1989b). Mathematics Problem Book: Fractions and Simple Algebra Dunedin, N.Z.: Dunedin Institute for Learning

Firestone, A.H. (1989c). All You Need to Know About Fractions and Simple Algebraic Equations Dunedin, N.Z.: Dunedin Institute for Learning.

Firestone, A.H. (1990a). A New Approach to Teaching Mathematics. Unpublished M.A. Thesis University of Otago, Dunedin (1990).

Firestone, A.H. (1990b). A Metacognitive, Rule-based, Conceptual, Problem-solving Approach to Teaching Fractions and Simple Algebra. Paper presented at 12th N.Z.A.R.E. Conference, Auckland College of Education(9-12 Dec. 1990).

Firestone, A.H., Nepe, T., Graham, J.D., & Glynn, T. (1990) Gains in Mathematics Accuracy Understanding and Confidence by Kura Kaupapa Maori Teacher Trainees Using 'Advance and Justify' Procedure. Final Report to Ministry of Education, Research and Statistics Dept. (June, 1990).

Firestone, A.H., Nepe, T., Graham, J.D., & Glynn, T. (1991) Gains in Mathematics Accuracy Understanding and Confidence by Kura Kaupapa Maori Teacher Trainees Using 'Advance and Justify' Procedure. Paper presented at 13th N.Z.A.R.E. Conference, Knox College, Dunedin (Dec. 1991).

Firestone, A.H. (1994, 24-26 August). Advance and Justify: a Metacognitive, Conceptual, Rule-based Approach to Learning. Paper presented at the Co-operative Education Asia- Pacific Conference (24-26 August 1994) Auckland, N.Z.

Flavell, J.H. (1976). Metacognitive Aspects of Problem-solving. In L.R. Resnick (Ed) The Nature of Intelligence (pp. 231 – 235). Hillside. New Jersey

Gagne, R.M. (1984). Useful Categories of Human Performance. American Psychologist . Vol. 39, 4, 377-385.

Glaserfeldt. E. (1989) Cognition, Construction of Knowledge and Teaching [10.1007/BF00869951]. Synthesis, 80 (1), 121140.

Gray. E.T.. Davis (1992). Success and Failure in Mathematics: The flexible meaning of symbols as Process and Concept. Mathematics Teaching, 142. Pg 6 -10)

Greeno, J. (1980). Trends in the Theory of Knowledge for Problem Solving. (Eds.) D. Tuma & F. Reif. Problem Solving and Education: Issues in Teaching and Research.

Hart, K.M.; Brown, M.L.; Kuchemann, D.E.; Kerslake, D.; Ruddock, G.; & McCartney, M. (1981). Children's Understanding of Mathematics: 11-16. London: Murray.

Hawks Bay Maths Advisors (1986). Hawkes Bay Math Series . Hawkes Bay: Maths Advisors, Department of Education.

Hayes, J.R. and Simon, H.A. (1980). Psychological Differences Among Problem Isomorphs. In N.J. Castellan Jr., D.B. Pisoni, and G.R. Potts (Eds.) Cognitive Theory (Vol. 2). Hillsdale, N.J.: Erlbaum.

Karpov, Yurly V. (2005). The neo-Vygotskian Approach to Child Development. N.Y.: Cambridge

Krutetskii, V.A. (1976). The Psychology of Mathematical Abilities in School Children. Chicago: The University Press. Lester, F.K. (1978). Mathematical Problem Solving in the Elementary School: Some Educational and Psychological Considerations. In Mathematical Problem Solving: Papers from a Research Workshop. Columbus, Ohio: ERIC Center for Science.

Lester, F.K. (1978). Mathematical Problem Solving in the Elementary School: Some Educational and Psychological Considerations. In Mathematical Problem Solving: Papers from a Research Workshop. Columbus, Ohio: ERIC Center for Science.

Lester, F, Jr. (1985). Methodological Considerations In Research on Mathematical Problem-Solving Instruction. In E. A. Silver (Ed.) Teaching and learning mathematical problem solving: Multiple Research Perspectives. Hillsdale, N.J.: Lawrence Erlbaum.

Mason, J., Burton, L., and Stacey K. (1982). Thinking Mathematically. London: Addison Wesley.

Matlin, M.W. (1989). Cognition. N.Y.: Holt, Rinehart, and Winston. Newel, A. and Simon H. (1972). Human Problem Solving. N.J.: Prentice Hall.

Polya, G. (1984 Vol. 2). Mathematical Discovery: On Understanding, Learning, and Teaching Problem Solving. London: John Wiley and Sons.

Resnick, L.B. (1983). Mathematics and Science Learning: A New Conception. Science, 220, 477-78, 1983.

Schoenfeld, A.H. (1979). Explicit Heuristics Training as a Variable in Problem Solving Performance. Journal for Research in Mathematics Education. 10 (3), 173-187.

Shultze, A.(1927). The Teaching of Mathematics in Secondary Schools. Hillsdale, N.J.: Lawrence Erlbaum Assoc.

Skemp, R.R. (1987). Psychology of Learning Mathematics. Hillsdale N.J.: Lawrence Erlbaum Assoc.

Vygotsky, L.S. (1962). Thought and Language. Cambridge, Mass.: MIT Press.

Biography

My name is Alex Firestone. Born in England, I emigrated to California with my family at age 10. I earned a Bachelor's degree in Mathematics and Science in Los Angeles, intending to become a teacher. However, I began my career as a research physicist, a role I held for four years before seeking a new challenge.

After moving to Australia with my wife, I taught Mathematics and Physics at both secondary and tertiary levels. Four years later, family matters brought us back to the U.S., where I resumed research. Following the birth of our fourth child, we relocated to New Zealand, but increasing rigidity in education made it difficult to teach mathematics the way I believed it should be taught. When three of our six children were diagnosed with learning difficulties, I left my role as Head of Mathematics, and we bought a neglected dairy farm. Working on the farm provided our children with a supportive learning environment.

These experiences led me to pursue a Master of Education, where my research demonstrated the success of my mathematics teaching methods. After earning my degree, I returned to teaching and founded my own school. I presented my findings at international conferences, but only China showed real interest. In 1985, I was invited to Chengdu to demonstrate my methods, but the country was not yet ready to implement them. Five years later, I was invited back, eventually leading to a PhD opportunity in Macau.

While lecturing part-time and pursuing my PhD at a Portuguese university in Macau, the 2008 economic downturn forced the cancellation of all foreign work visas, and I had to leave China. I continued my PhD studies at Griffith University, conducting workshops for pre-service teachers. My breakthrough came upon discovering Vygotsky's theories, which aligned with my own. Confident in my findings, I left my PhD program and moved to Cairns, securing a permanent Mathematics teaching position.

Unfortunately, severe health issues forced me to resign. The bright side was that I could finally focus on completing the book I had started in the 1980s. After 40 years, I have finished my book, bridging my mathematical insights with word-processing software—a long, challenging, but fulfilling journey.

www.ingramcontent.com/pod-product-compliance
Lightning Source LLC
Chambersburg PA
CBHW041004210326
41597CB00001B/5